Geotechnical Engineering - Advances in Soil Mechanics and Foundation Engineering

Edited by Sayed Hemeda and Mehmet Barış Can Ülker

Published in London, United Kingdom

IntechOpen

Supporting open minds since 2005

Geotechnical Engineering - Advances in Soil Mechanics and Foundation Engineering
http://dx.doi.org/10.5772/intechopen.78196
Edited by Sayed Hemeda and Mehmet Barış Can Ülker

Contributors
Xirong Niu, Yangsheng Ye, Qianli Zhang, Jinfei Chai, Xiaoyan Du, Hany Elosta, Fuping Gao, Wengang Qi, Mehmet Barış Can Barış Can Ülker, Kamil B. Afacan, Sayed Hemeda

First published in London, United Kingdom, 2020 by IntechOpen
IntechOpen is the global imprint of INTECHOPEN LIMITED, registered in England and Wales, registration number: 11086078, 7th floor, 10 Lower Thames Street, London, EC3R 6AF, United Kingdom
Printed in Croatia

British Library Cataloguing-in-Publication Data
A catalogue record for this book is available from the British Library

Additional hard and PDF copies can be obtained from orders@intechopen.com

Geotechnical Engineering - Advances in Soil Mechanics and Foundation Engineering
Edited by Sayed Hemeda and Mehmet Barış Can Ülker
p. cm.
Print ISBN 978-1-78984-289-0
Online ISBN 978-1-78984-290-6
eBook (PDF) ISBN 978-1-78985-302-5

For EU product safety concerns:
IN TECH d.o.o., Prolaz Marije Krucifikse Kozulić 3, 51000 Rijeka, Croatia, info@intechopen.com or visit our website at intechopen.com.

Meet the editors

Dr. Sayed Hemeda is a Doctor of Civil Engineering. He is currently Professor of Geotechnical Engineering and Architectural Preservation of Architectural Heritage, Conservation Department, Faculty of Archaeology, Cairo University, Egypt. He is also the manager of historic buildings Conservation Center in Cairo University. He is the recipient of many awards from Cairo University including prizes for scientific excellence (2017), encouragement (2014), and best PhD thesis (2009-2010). He was also awarded the General Union of Arab Archaeologists prize for academic excellence. He has published 85 articles, 23 books, and has been cited 130 times. He has given more than 38 invited lectures in 16 countries. His primary interests are geotechnical engineering for architectural heritage preservation and engineering data analysis including pattern recognition as applied to primarily analytical data from various sources such as objects of cultural significance. He is editor in chief for the *Journal of Geological Research*. He is an editorial board member for many organizations and publications including Springer Nature, IntechOpen, the *Open Journal of Geology*, *Progress of Electrical and Electronic Engineering*, *Geoscience Journal*, and *Alexandria Engineering Journal*.

Dr. M.B.C. Ülker graduated from the North Carolina State University Department of Civil, Construction and Environmental Engineering with a PhD degree in Civil Engineering in 2009. His doctoral research topic was on the dynamic analysis of saturated and nearly saturated porous media with applications to wave-induced seabed response. Following his PhD study, Dr. Ulker worked as an instructor at the Missouri University of Science and Technology Department of Civil Engineering in 2010. He then joined the Istanbul Technical University (ITU) Earthquake Engineering and Disaster Management Institute in 2012 and has been a professor at ITU since then. Dr. Ulker's research interests focus mainly on analytical and computational geomechanics with applications to geotechnical earthquake engineering, coastal and offshore geotechnical engineering and unsaturated soil mechanics. He is currently running an externally funded research group at ITU undertaking analytical, numerical, and experimental studies in these fields.

Contents

Preface

This book will be of interest to practical geotechnical engineers. It discusses contemporary issues related to soil mechanics and foundation engineering in earthwork projects, which are critical components in civil construction and often require detailed management techniques and unique solution methods to address failures and remedial measures. The geotechnical engineering community continues to find improved testing techniques for determining the sensitive properties of soil and rock, including stress wave-based non-destructive testing methods, as well as improvements to shallow and deep foundation design. To minimize failure during construction, contemporary issues and data may reveal useful lessons and information to improve project construction management and minimize economic losses. This book discusses these aspects using appropriate methods in a simple way.

It also discusses many interesting topics in geotechnical engineering such as modern geotechnical practice, geotechnical earthquake engineering, principals and practice in foundation design, slope stability analysis, modeling in geomechanics, and geotechnical engineering preservation of historical buildings and archeological sites.

This book brings together a small collection of chapters and covers a wide range of geotechnical problems and solutions. The first chapter is the introductory chapter. The second chapter is concerned with uniaxial and triaxial creep performance of calcarenitic and sandy oolitic limestone formations for stability analysis of Roman rock-cut tombs in Alexandria, Egypt. This work provides a geological map and geotechnical evaluation of the rock mass of these catacombs and their surroundings. We estimated the rock mass quality of the different members within the Alexandria calcarenitic formations using the long-term mechanical testing and rock mass rating and geological strength index systems. The third chapter deals with geotechnical response models for a steel compliant riser in deep water clays. The fourth chapter discusses in detail local scour around a monopile foundation for offshore wind turbines and scour effects on structural responses. Whereas the second, third, and fourth chapters treat the failures of materials and structures with deterministic methods, a frost heave deformation analysis model for a microheave filler is explained in the fifth chapter. Estimation of the excess pore pressure generation and nonlinear site response of liquefied areas is presented in the sixth chapter. Weathered granite soils are discussed in the seventh chapter.

I would like to express my gratitude to the publishers (IntechOpen) for their efforts in publishing this book and to the authors of the accepted manuscripts for their work and patience.

Prof. Dr Sayed Hemeda
Professor in Geotechnical Engineering and
Architectural Preservation of Historic Buildings,
Faculty of Archaeology,
Cairo University,
Giza, Egypt

Mehmet Barış Can Ülker, PhD.
Assoc. Prof. in Disaster Management Institute Earthquake Engineering Program,
Istanbul Technical University,
Istanbul, Turkey

Chapter 1

Introductory Chapter: Geotechnical Engineering in a Broad Perspective - New Advances in Emerging Fields

Mehmet Barış Can Ülker

1. New advances in geotechnical engineering

Geotechnical engineering is the branch of civil engineering that studies soils, structures, or structural components residing on or inside these soils. The mechanics of soils provides the necessary physical attributes to be used to understand the mechanisms that play a key role in what happens to soils under external effects, such as those induced by structural loads or changes in the water content. Thus, soil mechanics is mainly interested in the deformation characteristics of soils such that failure loads can be predicted more accurately. Therefore, it is more convenient and perhaps more accurate to think of the problem as a soil-structure system which is composed of a porous soil medium (as a soil profile with number of layers) and a foundation or another geotechnical engineering structure. As for the design of such soil-structure systems, settlement and the ultimate bearing capacity become the two major criteria that must be satisfied to withstand any external static or dynamic load. Once such stability criteria are set, it is expected that any soil-structure system achieves these requirements during the lifetime of the project.

Geotechnical engineering consists of the collection of subfields in a wide range of practical problems related to soils, rocks, slopes, foundations, walls, etc. While the principles of soil mechanics is the ultimate cookbook, recipe for each problem changes, and the engineers who are the chefs of the practice must be able to deduce the most edible and desired cuisine (i.e., engineering solution) employing their ultimate material, mathematics. As the current technology is advancing, it is more and more common that such engineering solutions follow the emerging technology that can be incorporated into geotechnical engineering research and practice to develop the most efficient, most viable, and most accurate solution. In order to achieve such an elaborate task, we, as geotechnical engineers, still make use of three tools at our disposal: (i) experimental methods, (ii) numerical methods, and (iii) analytical methods.

In this introductory chapter, three emerging fields of geotechnical engineering where there have been significant developments in the last couple of decades are touched upon. Those are the (i) numerical methods in geotechnical engineering, (ii) unsaturated soils, and (iii) offshore geotechnics. A short summary of the basic notion of what each field mainly investigates and what has been recently done in each topic in recent years is also given.

1.1 Numerical methods in geotechnical engineering

Numerical methods provide approximate solutions to complex engineering problems. Soils are heterogeneous, anisotropic, and multiphase particulate materials with nonlinear stress-strain relationships. Thus, any geotechnical engineering problem containing soils as a material or as an entire domain exhibits a certain level of complexity demanding robust numerical methods to tackle the issue. Such complexity is often associated with material behavior or boundary conditions or both. As it is obvious that no analytical solutions are available to such intricate problems, numerical methods are found to be quite handy and surely the only available option that yields approximate solutions with an acceptable error margin. Some of the commonly used numerical methods today are finite element method (FEM), finite difference method (FDM), boundary element method (BEM), discrete element method (DEM), finite volume method (FVM), material point method (MPM), hybrid methods, etc. While it is no straightforward task to decide what method to use, for most of the cases, the numerical method to employ is mostly problem-dependent. That is, the desired unknowns in the problem and the crucial details of what variables need to be calculated, where they are to be calculated and to what accuracy is wanted, will actually determine the most appropriate method.

In a brief summary of how numerical methods are employed, it should be noted that in most of these methods, the physical problem is first defined in the spatial domain, and if the problem is time-dependent, a temporal domain exists as well. Then the governing equations are written in these domains (or only along the boundaries, see BEM) in terms of the field variables which are simply the unknowns in the problem. In geotechnical engineering problems, field variables are mostly the displacements of the solid skeleton and the pore pressures in the voids of soils, that is, pore water or pore air pressures depending on whether the soil is modeled as a two-phase or a three-phase medium, respectively. Then the governing equations are either discretized in existing domains or simply defined at specified locations in the soil material, and the necessary boundary and initial conditions are prescribed in terms of field variables (or the degrees of freedom) and their derivatives with respect to Cartesian coordinates and/or to time. When the discretized forms of the governing equations are obtained, they are to be solved using an available numerical method for each load or time step. While such a clear-cut process is sufficient to provide some sort of a numerical solution to the problem, it is important to note that the ultimate solution will only be as accurate as how much the problem is simplified in the beginning of the process. Once the field variables are calculated, stresses (or any other stress-state variable) are computed at material points (i.e., convergence or integration points) in the soil elemental level through constitutive models using the strain field. Strains are essentially calculated utilizing the kinematic strain-displacement relations. For the soil skeleton, constitutive law is the effective stress-strain relationship, and for the soil pores, the necessary constitutive equation is the Darcy's law governing the flow of pore fluid due to differences in the total head between two points of interest in the soil.

Some of the trending problems encountered in geotechnical engineering in recent years, where numerical analysis is the key element, are:

- Constitutive theories for transient soil response. Here, particularly the theoretical models developed for predicting the cyclic behavior of soft soils and impact-induced response of special soils and rocks entail certain issues. Such a behavior can be physically modeled on a cyclic triaxial apparatus and numerically simulated in [1] (see **Figure 1**).

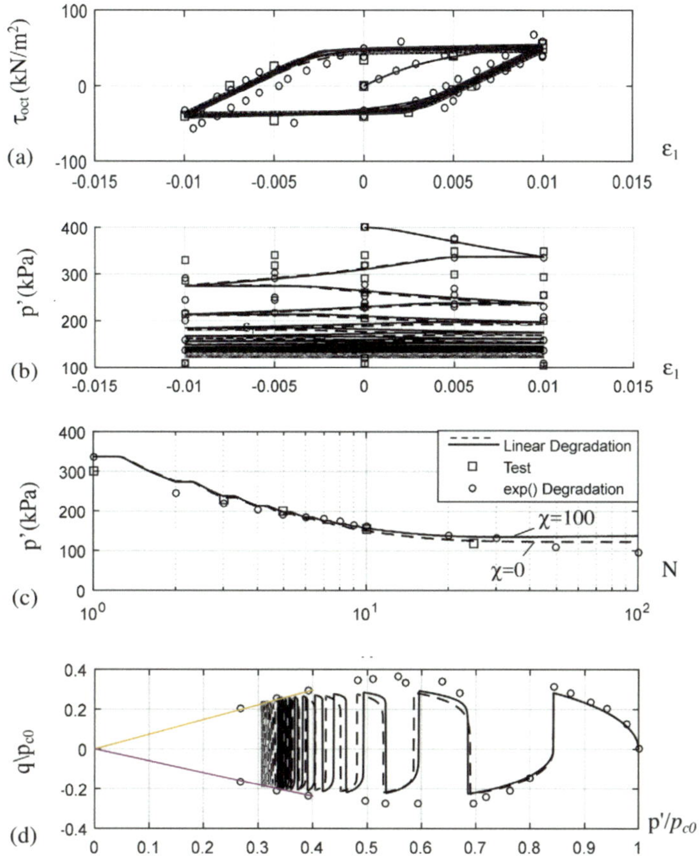

Figure 1.
Two-way strain-controlled undrained triaxial test simulation of kaolin clay as compared with [2] Grenoble workshop tests and [3]. (a) Shear stress-axial strain relationship. (b) Decreasing mean effective stress with axial strain of $\varepsilon_1 = \pm 1\%$. (c) Mean effective stress variation. (d) Stress path.

- Large deformation or post-failure behavior of soft soils. A serious issue related to soft ground is "post-liquefaction behavior" as well as settlements and lateral spreading of loosely deposited saturated sands subject to seismic motions. Another one is the pile-driving problem or modeling the penetration of piles into soil deposits. This is a typical example of a large deformation and post-failure behavior, which may pose a challenge in terms of numerical modelling.

- Uncertainties in distributions/initial conditions of mechanical quantities of soils. Geotechnical problems always suffer from lack of knowledge on soil properties and initial conditions. Inadequate and insufficient information in geotechnical problems frequently lead to a discrepancy between prediction and measurement, which implies that the prediction in these problems is accompanied with several types of uncertainties. While deterministic approach has since been quite useful in providing an estimate solution provided that there is certain amount of data assimilation and inverse analyses back the approach, probabilistic approach is also promising. On the latter approach, stochastic analysis is a technique that considers the uncertainty as a probabilistic variable. It is employed to assess the possibility of failure or other limit states of soil-structure systems. Probabilistic approach is directly related to reliability analysis and performance-based design.

1.2 Unsaturated soils

Another emerging field in geotechnical engineering is related to unsaturated soils. The response of unsaturated soils (USS) constitutes an important consideration for many problems in geotechnical engineering and geomechanics. Current geotechnical practice that models unsaturated soil mechanical properties as if they represent fully saturated conditions results in unsafe design due to a potential rise in the water content of soil. On the other hand, saturating natural soil samples in the laboratory, which normally would represent partially saturated state with in situ water contents, softens the soil which might lead to overdesign. Since such soil layers in the field are anticipated to have varying degrees of saturation during seasonal changes, this elevates the associated risks.

The developments in the field of “unsaturated soil mechanics” have lagged behind those of “saturated soil mechanics,” and the study of unsaturated soils has only recently gained importance. Although there have been many studies conducted in the last 30 years, there still needs experimental and theoretical works to be done to understand the USS hydromechanical behavior. Upon laboratory testing, predicted results should be calibrated such that the developed model could subsequently be employed to solve a practical problem through available numerical methods. This requires the model to be implemented into a computer program in terms of numerical algorithms developed for an elemental soil using USS parametric relations in a multiparametric space. This means, the development in the USS mechanics is in direct relation with the developments in the previous topic. On the one hand, there is the objective of eliminating the downsides of treating USS as fully saturated and therefore reducing the associated unconservative design owing to unpredicted behavior of USS. On the other hand, there is the uneconomical design to be reduced due to overconservative approach.

Water content in unsaturated soils plays an important role in evaluating the hydromechanical response in relation to suction. Behaviors of USS under static loadings are affected significantly by their volume changes as a function of the change in their water content. This is, however, directly related to the difference in the pressures of pore water and pore air which is called “matric suction,” s. The major stress variable controlling the deformation characteristics of USS is suction. As the matric suction increases, stiffness and shear strength of USS increase as well under constant net mean stress. As for the stress state, while a single effective stress is sufficient to describe the stress-strain relationship of saturated soils, this is not the case for USS. Hence, effective stress equation needs to be modified in formulating a constitutive model for USS. This issue has been the main subject of debate among the researchers since the pioneering study of [4]. Bishop [5] is the first to propose a relation for effective stress that accounts for the air phase on the average stress acting on the solid skeleton. The “Bishop stress” is written as the modified effective stress:

$$\sigma'' = \sigma - \chi s \tag{1}$$

where χ is called the Bishop parameter which is a function of the water degree of saturation. While some researchers claim that there is a “smooth” transition to the new definition of the effective stress [6–9], others say that there needs to be two independent stress variables, that is, the net stress, σ_{net} or $\overline{\sigma}$, and matric suction to govern the response of USS [4, 10, 11]. Net stress is defined as:

$$\overline{\sigma} = \sigma - u_a \tag{2}$$

It is important that in developing a constitutive model that accounts for hydro-mechanical behavior of USS, *hydraulic hysteresis* observed in the water retention behavior is taken into account. Such a behavior is self-evident in a typical constitutive relationship called the "soil-water characteristic curve (SWCC) or water retention curve (WRC)" irrespective of the soil type. During a drying-wetting cycle causing changes in saturation and suction, USS exhibits such hysteretic behavior. Therefore, it is necessary that the hysteretic response is incorporated in the mathematical formulation of a constitutive model for USS. Some of the common relationships are given in **Table 1**. There, θ is the volumetric water content, θ_s is the saturated water content (%), θ_r is the residual volumetric water content, and Ψ is the suction head with parameters, λ, a, m, and n, being empirical constants. **Figure 2** shows the data obtained by [12] using the HYPROP device which fits the curve by [13].

Recent elastoplastic models incorporating the hysteresis effect are distinguished in twofold: (i) models that account for hydraulic hysteresis through defining more yield surfaces [7, 16, 17] and (ii) models in which the effect of hydraulic hysteresis is included in the evolution of a so-called *load collapse*, (LC) curve [6, 18]. LC curve defines the evolution of "pre-consolidation pressure," p_0, of the soil with increasing plastic strains. Also, there are two other yield curves controlling the yielding of USS as matric suction changes called the *suction increase* (SI) and *suction decrease* (SD) yield curves (i.e., [16, 19]).

Model	Relation
[14]	$\theta(\Psi) = \theta_r + (\theta_s - \theta_r)(\alpha\Psi)^{-\lambda}$
[13]	$\theta(\Psi) = \theta_r + \frac{\theta_s - \theta_r}{\{ln\,[e + (a\Psi)^n]^m\}}$
[15] ($m = 1 - 1/n$)	$\theta(\Psi) = \theta_r + (\theta_s - \theta_r)[1 + (\alpha\Psi)^n]^{-m}$
[15] (m, n independent)	$\theta(\Psi) = \theta_r + (\theta_s - \theta_r)[1 + (\alpha\Psi)^n]^{-m}$

Table 1.
Some commonly used SWCC relations.

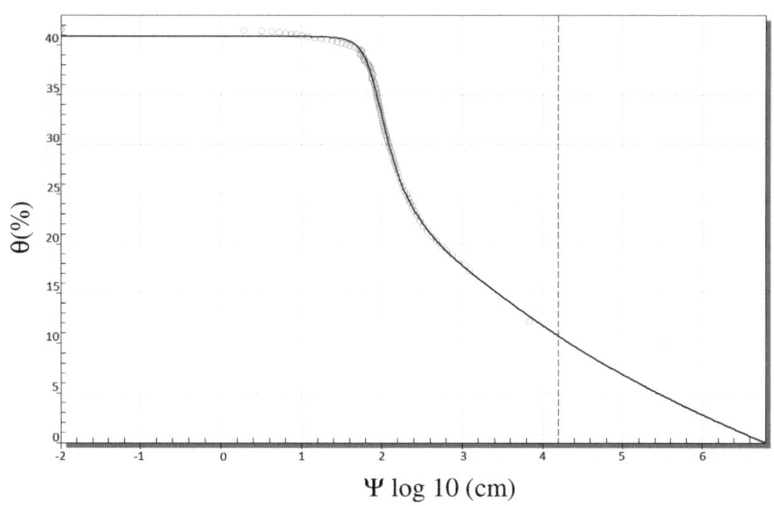

Figure 2.
SWCC of Mersin silt fitted to Fredlund and Xing [13] (after [12]).

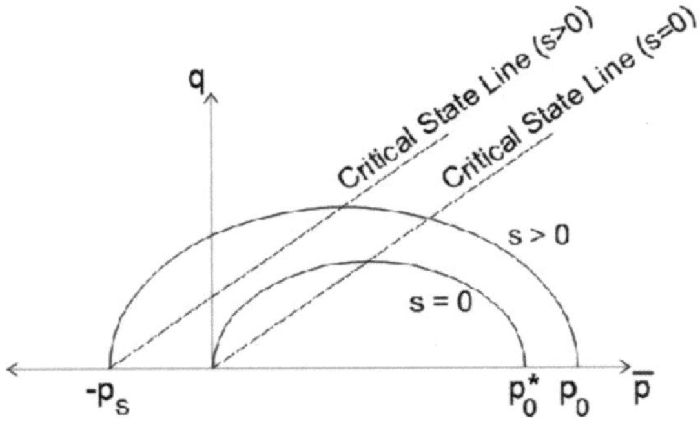

Figure 3.
Dependence of the yield surface on suction.

Barcelona Basic Model (BBM) [4] is considered to be the pioneering study that explicitly proposes an elastoplastic framework to understand the hydromechanical behavior of USS. The model is based upon the critical state theory with a classical plasticity framework employing a nonassociated flow rule. BBM adopts the yield surface of Modified Cam-Clay (MCC) model in the constant suction surface (Eq. (3)), whose size depends on suction (**Figure 3**). The yield function is written in triaxial stress components as:

$$q^2 - M^2(\overline{p} + p_s)(p_0 - \overline{p}) = 0 \tag{3}$$

As far as hydromechanical coupling, generally speaking, the yield criterion should consist of a hardening parameter that is a function of both suction and plastic strain, since stiffness and strength characteristics of unsaturated soils increase with increasing values of hardening parameters. Thus we have:

$$f(\overline{\sigma}, h) = 0 \tag{4}$$

$$h = h(\varepsilon^p, s) \tag{5}$$

where h is the "hardening parameter." In BBM per (3), $h = p_0$ emphasizes that the consistency condition written for the yield surface must include suction as well as net mean stress. Consistency condition for the yield surface is then written as:

$$df = \frac{\partial f}{\partial \overline{\sigma}}^T d\overline{\sigma} + \frac{\partial f}{\partial p_0}\frac{\partial p_0}{\partial \varepsilon^p}^T d\varepsilon^p + \frac{\partial f}{\partial p_0}\frac{\partial p_0}{\partial s}^T ds + \frac{\partial f}{\partial p_s}\frac{\partial p_s}{\partial s} ds = 0 \tag{6}$$

While the formulation of the mechanical part of USS continues with the steps of classical plasticity starting from (Eq. (6)), the hydraulic part essentially starts off with the partial derivative of volumetric water content written as:

$$d\theta = S_r * dn + n * dS_r \tag{7}$$

where S_r is the degree of saturation and n is porosity. This equation can be expanded to yield a form employing matric suction and volumetric strain [20]. Derivation of S_r with respect to suction can then be obtained using the water retention behavior in terms of the soil-water characteristic curve. The final form of the coupling relation is obtained as:

$$d\overline{\sigma} = D_{hm}^{ep}\, d\varepsilon_{hm} \tag{8}$$

where $D^{ep}{}_{hm}$ is the *hydromechanical elastoplastic stiffness matrix*. An example of predicted behavior of USS can be seen in **Figure 4** as compared to the BBM.

In summary, factors controlling the USS response are complicated due to soils' particulate, multiphase, and nonlinear nature. Particularly, the volume-saturation relation and water retention behavior of USS, unlike saturated soils, provide additional complexity due to existence of matric suction. Hydraulic hysteresis is another response of USS observed in water retention behavior regardless of the soil type. Therefore, a constitutive model developed to understand the hydromechanical response of unsaturated soils has to account for these facts observed in specialized laboratory tests.

1.3 Offshore geotechnics

The last field of interest that gains popularity mainly in the last two decades is offshore geotechnical engineering. Offshore structures that reside on or around seabed soil are essential for energy, protection, or transmission. The stability of the whole soil-structure system relies not only on the structural integrity under wave action but that the seabed soil must be able to withstand the induced stresses and pore pressure buildup against various loadings due to wave and currents. Geotechnical considerations are important in identifying such conditions leading to instability of seabed-offshore structure systems under wave action. Therefore, it is important that wave-induced soil behavior is modeled via appropriate theoretical frameworks.

The developments in this field depend upon three interrelated subtopics: (i) wave mechanics, (ii) structural mechanics, and (iii) soil mechanics. Hence, the fluid-soil-structure interaction (FSSI) is an indispensable field of study that one who is working on offshore geotechnics needs to tackle with diligence. In addition, the studies made in this field are also related to the topics discussed in the previous sections in that numerical methods are frequently employed to provide approximate engineering solutions to related problems. Also, while it is not common to observe unsaturated soils in offshore environments (except the air voids made up due to sea shells), aspects of soil constitutive modeling associated with unsaturated soils are indirectly applicable to the elemental-level response of saturated soils encountered in seabed soils. Thus, in this section of this introductory chapter, the focus is on such relationships between seabed soils and their constitutive behavior under cyclic wave loading as opposed to specific offshore structures and their analysis details. Interested readers can refer to the recent proceedings published by the International Society of Offshore and Polar Engineers (ISOPE) [21–23].

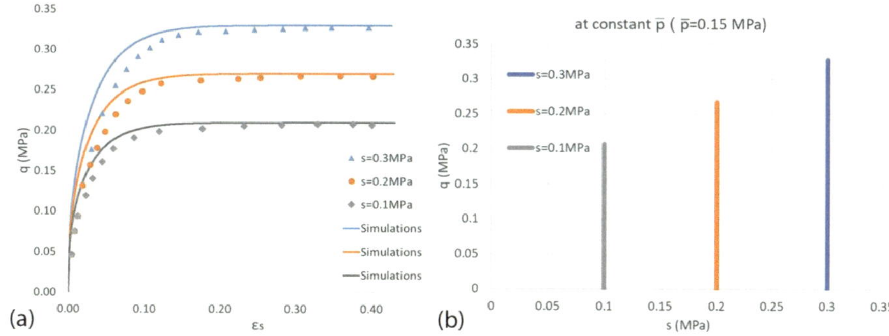

Figure 4.
Prediction of hydromechanics of USS. (a) Verification with BBM [4] and (b) stress paths (after [20]).

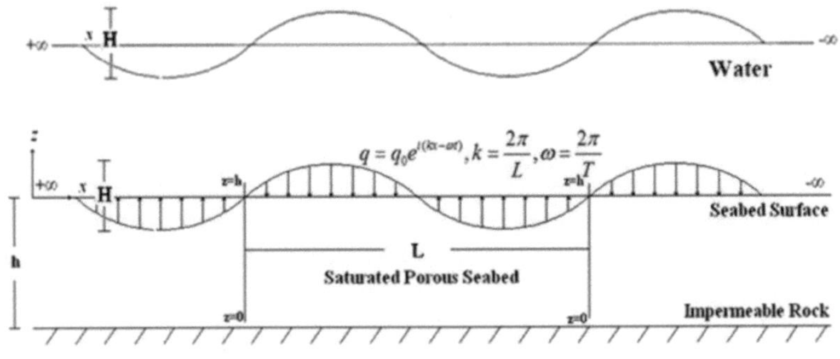

Figure 5.
Free field wave-induced saturated porous seabed response.

Figure 5 shows the progressive wave-induced seabed problem in the free field. Here, wave loading induces a similar harmonic stress variation on a localized seabed layer. The goal is to calculate the displacements and pore pressures in the seabed under wave action. Surely the actual wave effect will have an irregular structure, but this is a sufficient representation of the wave-induced response of the system.

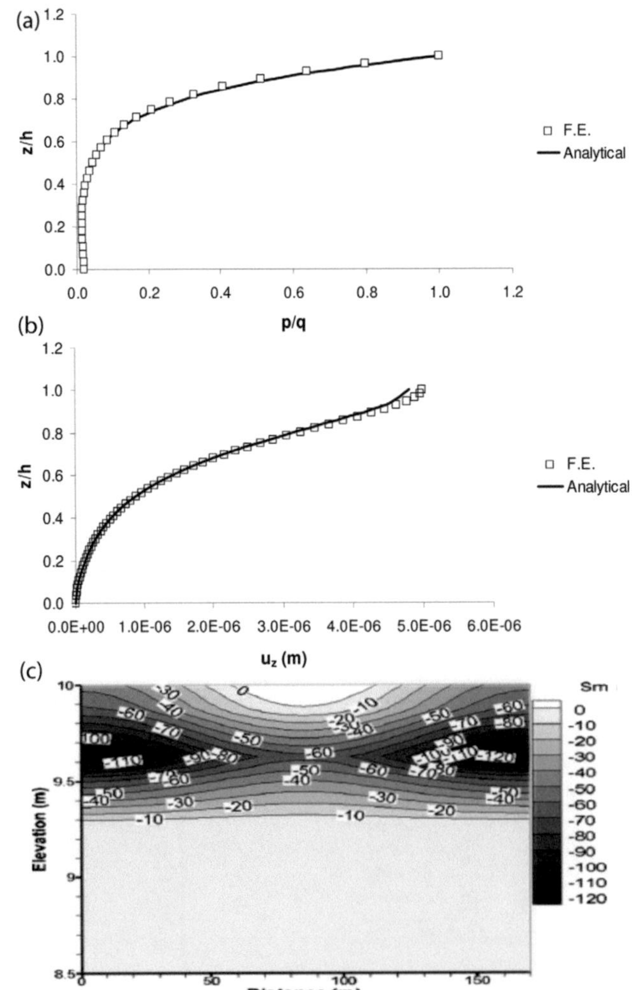

Figure 6.
(a) Pore pressure, (b) displacement distribution and (c) liquefaction response within the seabed.

The solution of this problem using analytical and numerical means where the equations governing the dynamic response of seabed soil are that of coupled flow and deformation of Biot poroelasticity [24, 25] is given in **Figure 6a**, **b** [26, 27]. Prior to employing any numerical methods in developing an associated formulation, it is highly recommended that the researchers dealt with the free field response problem so that they have a better understanding of the elemental behavior of saturated porous seabed. In addition, in this way the wave-induced instantaneous liquefaction phenomenon observed in seabed based on mean effective stress (Sm) criterion can also be modeled [28] (**Figure 6c**). Interested readers for more on the subject matter in this field can refer to the wide literature available on the topic, some of which can be found in [28–31].

Author details

Mehmet Barış Can Ülker
Institute of Disaster Management Earthquake Engineering Program, Istanbul, Turkey

*Address all correspondence to: mbulker@itu.edu.tr

References

[1] Ülker MBC. A new hardening interpolation rule for the dynamic behavior of soils using generalized plasticity framework. In: ICSMGE 2017—19th International Conference on Soil Mechanics and Geotechnical Engineering; 2017

[2] Kuntsche K. Tests on clay. In: Proc. International Workshop on Constitutive Behavior of Soils; Grenoble; 1982

[3] Zienkiewicz OC, Leung KH, Pastor M. Simple model for transient soil loading in earthquake analysis. I. Basic model and its application. International Journal for Numerical and Analytical Methods in Geomechanics. 1985;**9**:453-476. DOI: 10.1002/nag.1610090505

[4] Alonso EE, Gens A, Josa A. A constitutive model for partially saturated soils. Geotechnique. 1990;**40**: 405-430

[5] Bishop A. The principle of effective stress. Teknisk Ukeblad. 1959;**39**:859-863

[6] Gallipoli D, Gens A, Sharma R, Vaunat J. An elasto-plastic model for unsaturated soil incorporating the effects of suction and degree of saturation on mechanical behavior. Géotechnique. 2003;**53**:123-136

[7] Wheeler SJ, Sharma R, Buisson M. Coupling of hydraulic hysteresis and stress–strain behavior in unsaturated soils. Geotechnique. 2003;**53**:41-54

[8] Jommi C. Remarks on the constitutive modelling of unsaturated soils. In: Tarantino A, Mancuso C, editors. Experimental Evidence and Theoretical Approaches in Unsaturated Soils. Rotterdam: Taylor & Francis; 2000. pp. 121-138. ISBN: 90-5809-186-4

[9] Sun D, Sheng D, Xiang L, Sloan S. Elastoplastic prediction of hydro-mechanical behaviour of unsaturated soils under undrained conditions. Computers and Geotechnics. 2008;**35**:845-852

[10] Fredlund D, Morgenstern N. Stress state variables for unsaturated soils. Journal of Geotechnical and Geoenvironmental Engineering. 1977; **103**:447-466

[11] Fredlund DG, Pham HQ. Independent roles of the stress state variables on volume–mass constitutive relations. Theoretical and Numerical Unsaturated Soil Mechanics. Berlin, Heidelberg: Springer; 2007:37-44. DOI: 10.1007/3-540-69876-0_5

[12] Eyupgiller M, Ulker M. Determination of the relation between hydrolic conductivity and matric suction in silty soils with respect to saturation. In: 17th National Soil Mechanics and Geotechnical Engineering Conference; 2018

[13] Fredlund D, Xing A. Equations for the soil-water characteristic curve. Canadian Geotechnical Journal. 1994;**31**: 521-532

[14] Brooks AT, Corey RH. Hydraulic properties of porous media and their relation to drainage design. Transactions of ASAE. 1964;7:26-28

[15] Van Genuchten MT. A closed-form equation for predicting the hydraulic conductivity of unsaturated soils. Soil Science Society of America Journal. 1980;**44**:892-898

[16] Vaunat J, Romero E. An elastoplastic hydro-mechanical model for unsaturated soils. In: Tarantino A, Mancuso C, editors. Experimental Evidence and Theoretical Approaches in Unsaturated Soils. Rotterdam: Taylor & Francis; 2000. pp. 121-138. ISBN: 90 5809 186 4

[17] Gens A, Sheng D, Sloan SW. A constitutive model for unsaturated soils: Thermomechanical and computational aspects. Computational Mechanics. 2004:**33**:453-465

[18] Tamagnini R. An extended Cam-clay model for unsaturated soils with hydraulic hysteresis. Geotechnique. 2004;**54**:223-228

[19] Sheng D. Review of fundamental principles in modelling unsaturated soil behaviour. Computers and Geotechnics. 2011;**38**:757-776. DOI: 10.1016/j.compgeo.2011.05.002

[20] Eyupgiller M, Ulker MBC. Effect of plastic deviatoric strains on the hardening of unsaturated soils in relation to their hydromechanical behavior. In: 8th National Geotechnical Symposium; 2019

[21] Ulker MBC. Constitutive modeling of cyclic seabed behavior around coastal and offshore structures: Two robust models and their predicting capabilities. In: Proc. 29th International Ocean and Polar Engineering Conference; Honolulu, HI: ISOPE; 2019

[22] Baksi HG, Ülker MBC. Analysis of dynamic response of a caisson type gravity quay wall—Seabed soil system under waves. In: Proceedings of the International Offshore and Polar Engineering Conference; 2017

[23] Ülker MBC, Fard MM. Instantaneous liquefaction of seabed around rubble mound breakwater under waves. In: Proceedings of the International Offshore and Polar Engineering Conference; 2016

[24] Biot MA. General theory of three-dimensional consolidation. Journal of Applied Physics. 1941;**12**:155-164. DOI: 10.1063/1.1712886

[25] Biot MA. Theory of elasticity and consolidation for a porous anisotropic solid. Journal of Applied Physics. 1955; **26**:182-185. DOI: 10.1063/1.1721956

[26] Ulker MBC, Rahman MS. Response of saturated and nearly saturated porous media: Different formulations and their applicability. International Journal for Numerical and Analytical Methods in Geomechanics. 2009;**33**:633-664. DOI: 10.1002/nag.739

[27] Ulker MBC, Rahman MS, Jeng DS. Wave-induced response of seabed: Various formulations and their applicability. Applied Ocean Research. 2009;**31**:12-24. DOI: 10.1016/j.apor.2009.03.003

[28] Ulker MBC. Pore pressure, stress distributions, and instantaneous liquefaction of two-layer soil under waves. Journal of Waterway, Port, Coastal, and Ocean Engineering. 2012; **138**:435-450. DOI: 10.1061/(ASCE)WW.1943-5460.0000155

[29] Ülker MBC. Wave-induced dynamic response of saturated multi-layer porous media: Analytical solutions and validity regions of various formulations in non-dimensional parametric space. Soil Dynamics and Earthquake Engineering. 2014;**66**:352-367. DOI: 10.1016/j.soildyn.2014.08.005

[30] Ülker MBC. Instantaneous liquefaction in a multi-layer seabed under nonlinear progressive waves. AIP Conference Proceedings. 2012:1479-1484. DOI: 10.1063/1.4756444

[31] Ulker MBC, Tatlioglu E, Lav MA. Dynamic response and liquefaction analysis of seabed-rubble mound breakwater system under waves. Applied Ocean Research. 2018;**78**:75-87. DOI: 10.1016/j.apor.2018.06.008

Chapter 2

Uniaxial and Triaxial Creep Performance of Calcarenitic and Sandy Oolitic Limestone Formations for Stability Analysis of Roman Rock-Cut Tombs in Alexandria, Egypt

Sayed Hemeda

Abstract

The Greek-Roman rock-cut tombs at Alexandria, Egypt, were excavated mainly in the calcarenitic limestone formations and show varying degrees of damage of rock pillars and ceilings. In order to understand the long-term rock mass behaviour in selected tombs and its impact on past failures and current stability, uniaxial and triaxial Creep tests and rock mass quality assessments had been carried out. Creep behavior of rock plays an important role in underground works, especially for archeological structures subjected to large initial stresses. These conditions yield nonreversible deviatoric creep strains that develop during time at constant stress. In order to describe the time-dependent deformation, various approaches have been established based on analytical, empirical, and numerical methods. Our analyses show that the Roman tombs at Alexandria have been cut into poor quality rock masses. Rock failures of ceilings and pillars were frequently facilitated by local, unfavourably oriented persistent discontinuities, such as tension cracks and joints. Other failures were related to the disintegration of calcarenitic and oolitic limestones. Our data suggest that, in Roman age monumental tomb construction, low-strength rock masses resulted in modifications of the planned tomb design in order to minimise the risk of rock falls and to prevent collapses.

Keywords: Roman rock-cut tombs, geotechnical assessment, creep tests, calcarenitic limestone, oolitic limestone, rock mass rating, tomb construction

1. Introduction

Creep is an irreversible ductile time-dependent deformation, without fracture where deformation does not occur suddenly when applying stress as opposed to brittle fracture. Instead, strain accumulates as a result of long-term stress [1–4]. This behavior usually distinguishes weak rocks such as rock salt, shale, buttocks, venetian, silt, and sandstone. Rock creep behavior has been widely discussed in the literature, based on experimental results from laboratory or field investigations, foundational modeling, and numerical analyses. The main objective of this chapter is to focus on appropriate methodologies for determining the creep behavior of soft/weak intact rocks through laboratory experimental analysis and critical evaluation of available rheological models to explain creep behavior [5–9].

Creeping in fragile hard rocks is rare because the deformation rate is too slow. Solid rocks exhibit a creep behavior noticeably only at elevated temperatures and pressures generally not encountered in engineering structures. Soft rocks on the other hand mostly creep at room temperature, atmospheric pressure, and the range of deviating stress typically encountered in engineering structures [10–12].

As we know, creeping rocks have a significant effect on the long-term stability of the rocks and the surrounding surface [13–16]. For broken rocks, porosity is the primary determinant of creep characteristics, but in the existing literature, the stress rate was mainly used to describe the creeping properties of broken rocks. For example, Wang [17, 18] carried out numerical simulations on the process of creeping damage to the road surrounding the rock under high pressure, and Zhu and Ye discussed the law of creep affected by water content by comparing the results of the rock creep test under dry condition and in saturation. Zhang and Luo [19] studied the properties of creeping rocks under different stress levels. Liu et al. [20] performed triaxial creep tests on coal and rock by step loading method. Zhang and Luo [19] examined the creeping test of marble and soft rock separately; Parkin [21] used a pressure meter to study the rheological properties of granular materials. Shen and Zhao [22] proposed a model for three parameters of creeping rock filling through rheological experiments on limestone. Zheng and Ding proposed a creep model to rocks of nine parameters and obtained parameter indexes through tests. Guo et al. [23] proposed a modified three-parameter rheological model for coarse-grained materials. Wang [24] and Liu et al. [25] summarized the rheological state of coarse-grained materials and noted that experimental studies on granular materials were insufficient.

They suggested that it is necessary to study the mechanism of partial deformation of coarse granular materials given the effect of the scale for internal testing.

Understanding the mechanisms of deterioration of the calcarenite rock structures in which the Greco-Roman monuments are excavated requires a comprehensive study of the mechanical behavior and engineering properties of the calcarenite rocks. In addition to geological and geomorphological concerns, numerous investigations have been conducted on rock degradation and disintegration. As the areas are an open museum and attractive places for tourists, sampling can only take place in a limited number of locations with official permission. For this purpose, cylindrical samples with a diameter of 42–44 mm and a height of 90–100 mm, prepared using a basic drilling machine and some blocks collected from archeological sites under investigation (catacomb from Kom El-Shoqafa, El-Shatby tombs, and tombs of Mustafa Kamel), as shown in **Figure 1** illustrates the physical, short, and long-term mechanical properties of calcarenitic rocks in the laboratory, a number of samples prepared from these blocks have been used for testing, and the limitations of the number of blocks have been overcome by determining the topical properties of the rocks through hammer tests. Schmidt, pictorial geographic investigations and classification of the rocky hill in some outcrops and in some rock structures where testing was permitted.

The purpose of this research is to make recommendations on the strengthening and safety of archeological underground structures under long- and short-term loading. For this purpose, a set of experimental tests and advanced digital analyses had been performed.

Figure 1.
Underground monuments (Catacombs) in Alexandria, (present conditions).

Calcarenitic rocks and other type of fine limestone (under investigations) are porous rocks with complex behavior [26–28]. Two major mechanisms can be identified to distort types of rock properties, depending on conditions in-situ stress: (1) the prevalence of fracture, associated with volumetric expansion and fragile behavior, which is predominant in compressive stress paths in the absence of low confined pressure, or (2) pore breakdown, which dominates high-stress conditions, producing plastic deformations and large contracting [29].

The high fossil content, mainly due to the shells of necrosis and some mollusks, leads to structural heterogeneity, which is reflected in the variance of mechanical properties and weaknesses in the conclusion of experimental results [13].

There is no generally accepted theory of fragile rock strength based on examination of the process of formation of microcracks and deformation, and the establishment of the initiation and development of stress-induced fractures in EDZ is therefore a major concern.

Some of the main concerns related to the stability of underground structures in soft rocks include the effects of potential land disturbances through the method of drilling and reallocation of pressures at the site surrounding the excavations [30–35]. Each of these factors relates to the initiation and spread of fragile fractures and the extent of the troubled drilling area (EDZ), which can adversely affect the stability of the drilling boundaries and can increase the permeability of host rocks to the near field. In structural and tectonic geology, experimental rock deformation is important in determining the evolution of natural structures and tectonic features [36, 37].

Great effort has been made toward understanding the fragile fracture processes and mechanisms. Much of this focus extended to laboratory tests and quantification/measurement of fragile fracture thresholds [7, 38]. Among these, damaged thresholds marked by the onset of expansion, which is the reflection point of the volumetric pressure curve, are particularly important because many studies have linked the threshold to the spread of unstable fracture in fragile rocks [7]. The unstable crack spreading corresponds to the point where the reproduction process is controlled between the applied stress and the speed of crack growth. Under these circumstances, the crack will continue to spread until failure even if the applied load stops and remains stable. As such, Martin and Chandler and Read et al. equated the threshold of damage caused by cracking and the long-term on-site strength of fragile rocks.

Thus, the identification of these processes and associated mechanisms is essential in predicting both the strength of soft rocks in the short and long term. This research focuses on these processes by presenting the results of many short- and long-term laboratory tests.

In general, the spread of cracks can be equated with the irreversible destruction of molecular cohesion along the path of the crack generated. In this sense, the miniature crushing process “damages” the rock material. Due to the multiplication of the number of reproductive fractures, the damage can be considered to be cumulative and can be associated with a perceived lack of elastic stiffness and the strength of material cohesion.

In this work, we highlighted some important characteristics of the geotechnical behavior of structured soft rocks and showed that these properties are very common in many natural rocks. Based on these concepts, research into soil/rocky transition material has intensified in the last two decades [39].

2. Geological and tectonic setting

The Roman underground tombs in Alexandria are located on the northern edge of the “Nile Delta geomorphic province, c. 1.30 km north of the Lake Maryout and 1.42 km south of the Mediterranean Sea shoreline, as shown in **Figure 2**”. Since Pleistocene time, within the last 1 million years, Lake Maryout has intermittently

Figure 2.
The limestone outcrop at the catacombs of Kom El-Shoqafa.

been connected to fresh Nile river flows and sea water sources and has been both at and below mean sea level. Lake Maryout and Delta had varied depositional environment, including "silt and clay deposits with some organics (lagoonal deposit)"; "sand and silt deposits (Nile River deposits); "sand deposits (beach and littoral deposits"). The basement rock unit is Miocene (6–25 million years old) and older carbonate formations that comprise the Egyptian plateau. Above the Miocene sedimentary rocks are Plio-Pleistocene age (less than 6 million years old) sediments consisting of alternating beds of shale, limestone, sandstone, silt, and calcareous sand.

The Plio-Pleistocene sediments form a series of ridge and trough that are approximately parallel to the Mediterranean coastline in the vicinity of the catacomb site. Most of the city of Alexandria rests on one of these topographic ridges while behind the ridge, Lake Maryout is in a trough. The near surface limestone deposits, which are commonly encountered in the Alexandrian ridge, are cemented marine sand.

3. State of preservation

The catacombs of Alexandria show some clear indications of yield and partial collapse in several locations, as defined in the honeycomb weathering, the contour scaling and spalling of the stone surface, the disintegration of building materials, and the wet surfaces of rocky meals especially for semi-protected parts of the excavation; also, we observe salt flowering and yellow staining of yellow iron in many wall parts.

Structural damage is obvious like the wall cracking, the thinning out of rock pillars, disintegration and degradation of the walls surfaces, the partial collapse of some parts of the roofs and walls, and the peeling of rocks, especially in the roof of narrow corridors found in the deepest parts and mass waste from the ceiling and walls.

In conclusion, the current state of conservation of the great catacombs at Kom El-Shoqafa, the best-known and most famous testimony of the culture of the funerary architecture of Alexandria, is now at its most deteriorating.

Most structural damage is caused by one or a combination of the following factors:

- The gradual weakening of rock materials due to the intrinsic sensitivity of weathering factors, especially the effect of weathering with groundwater and salt
- Earthquake and other man-made dynamic loading
- Permanent deformation of the rock mass
- Natural wear and tear of materials
- History of construction in the area

4. Mineralogical and petrographical studies

The effort behind thin-section analysis was to provide insight into the closed grains (calcite/sand) and/or theories of overgrowth after precipitation of the large angle of internal friction. Due to the fragile nature of the rocks and plaster layers being excavated, it was necessary to be very careful to make thin sections, which were studied using independent polarized light, electron microscopy (SEM), and stereoscopic observation.

A light-transmitted polarized plane, scanning electron microscopy, and stereoscopic observations were used to determine the interlocking textures and connections between grains and crystals. These contacts rely on differences in solubility due to impurities and differences in bending radius, which lead to the penetration of smaller grains in large grains.

In addition, thin section microscopy was used to help explain the large friction angles associated with the material, limestone/rock.

Miniature petrographic description of stones/rocks for engineering purposes includes the identification of all parameters that cannot be obtained from a comprehensive endoscopic examination of rock samples, such as mineral content, grain size and texture, which have an impact on the mechanical behavior of the rock or rock mass. To ensure proper classification, the first step should be to check the metal composition and rock texture; see **Table 1**. Mineralogy summarizes the three types of soft limestone under investigation. Additional investigations should include analysis of the texture and minerals in the case of highly contrasting rocks, determining the degree of change or weathering, grain size, partial fracture, and porosity.

In sandstone, limestone, and calcarenite samples intact, it is possible to determine with the naked eye an alternative sequence of white and pink bands with a

Rock type	Calcite $CaCO_3$ %	Quartz SiO_2 %	Gypsum $CaSO_4.2H_2O$ %	Halite NaCl %	Other %
Sandy oolitic limestone (Kom El-Shoqafa) (COM)	47–65	31–23	10–5	12–9	2
Intact Calcarenite (Mustafa Kamel Necropolis) (M)	52–72	28–18	8–3	12–6	3–5
Oolitic intraclastic limestone (El-Shatby Necropolis) (SH)	53–80	25–20	11–7	11–7	3

Table 1.
Mineralogy of the three soft limestone types under investigation.

thickness of about 1 mm (bedding plane). Optical microscopy and counting points were performed on thin sections of rock samples. The air-dried samples were inoculated with Canada balsam, and the thin sections were then cut perpendicular to the bedding planes. A thin section is observed under parallel light and polarizing light. The following is a detailed analysis of the rock samples collected from the three archeological sites under investigation, rock samples from six collections of El-Shatby with code Nr (SH), five rock samples collected from the tombs of Mustafa Kamel 1 and No. 2 with code Nr (M), and four samples Rock collected from Catacomb of Kom El-Shoqafa code Nr (COM).

4.1 Catacombs of Kom El-Shoqafa

In the internal structure, we can observe the dominant components, which are the cells of the fibers of the stomach, grass, algae, and mother of pearl, mostly with a test wall of microscopic microspheres, while the tests are filled internally with microtomes and microbes (**Figure 3**). Surrounded monocrystalline quartz granules of varying sizes and perimeter of iron oxides have been detected representing the previous presence of K-feldspar grains. Rock and granular materials make up this fossil sand limestone, or cement sand.

(Calcarenite size) 15% of customizations are medium-sized numulite tests filled with prickly calcite. 15% of foraminifers tests with a neomorphic microspar test wall and test chambers are full of neomorphic microspar. 20% of medium size bryoza and algae tests 0.25% small size, monocrystalline, crispy extinction, quartz granules subrounded. 25% medium to small size structure less ooides. Customizations are solidified by isopachous microspar. Porosity is a fit of 20% of the area of the thin-section field, which is reduced by microscopy. Oxidation is observed as red color spots.

4.2 Mustafa Kamel Necropolis

The rock texture in these tombs consists of two textures, namely packed stone and stone. These two types of texture show different proportions and sizes of quartz granules, and different biological plates, especially foraminifer tests. Most Ooides lost their internal structure. Few of them retain their concentric structure. Consolidation of the components of this limestone is represented by isopachous microspar (**Figure 4**).

(Calc rud –arenite size) 58% of the assignments are medium in size, thin and micro pigment and less pollutant internal structure. 10% micritic oval. 30% large to small angular size to subrounded, crispy extinction, monocrystalline quartz. 2% plajioclase and microcline crystals. Porosity reached 20% of the area of the thin-section field. The pores are filled with neomorphic microspar. Allochems are surrounded with isopachous microspar.

Calcarenite is a bio-soft rock originating from marine sediments, which occurred during the overflow and decline of the region in the Ice Age. The calcarenite consists of almost pure calcium carbonate and is applied directly to the limestone rock of the Cretaceous.

4.3 El-Shatby cemetery

Changes in internal structure and metals were analyzed and the most distinctive textures documented on the images. In the internal structure, we can observe the porosity increase of various sizes. In some places, we can find cracks on metal contacts or even inside metals. Generally, significant changes are shown in the cement material; see **Figure 5**. Limestone in this site can be classified into two types

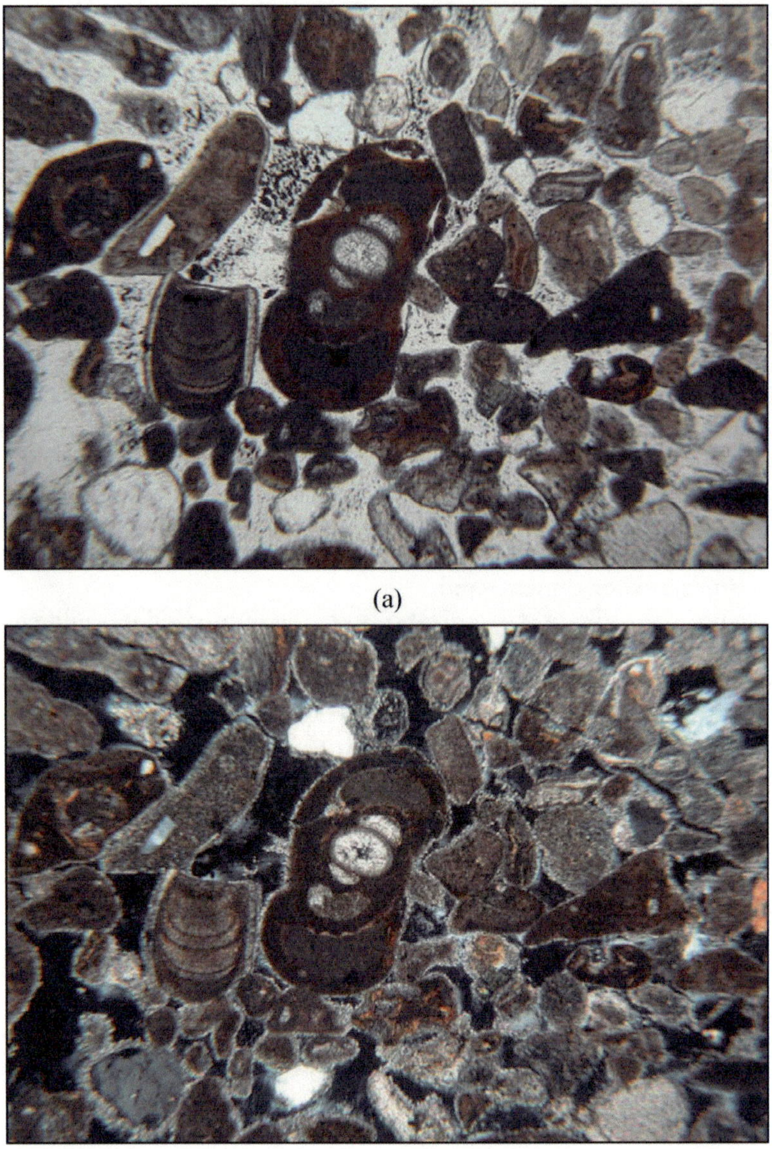

(a)

(b)

Figure 3.
Photomicrograph of fossiliferous sandy oolitic limestone, (a) under parallel polarized light, (b) under cross polarized light (XPL), showing bioclasts of gastropods, foraminifera, algae, and shell debris; most of them are with test wall of neomorphic microspar, filled with micrite and microspar, cracks between and through the minerals are obvious. Catacombs of Kom El-Shoqafa.

of fabric, namely, fossiliferous oolitic intraclasic limestone. These two types of texture are in different proportions of quartz granules, biological panels, ooides, and peloids.

(Callus arinite rod size) 80% of the customizations are medium-sized structure less ooides. 10% large to medium-sized monocrystalline unite extinction, quartz granules subrounded. 5% large polycrystalline, crispy extinction, quartz granules subrounded. Five% of algae and foraminifera are tested with a micritic wall and are filled internally with microscopic grains. Porosity is greatly reduced due to their filling with depressed dwarfs. The new form is observed to worsen from micrite to microspar. The evaluation of thin sections allows the analysis of pore structure and

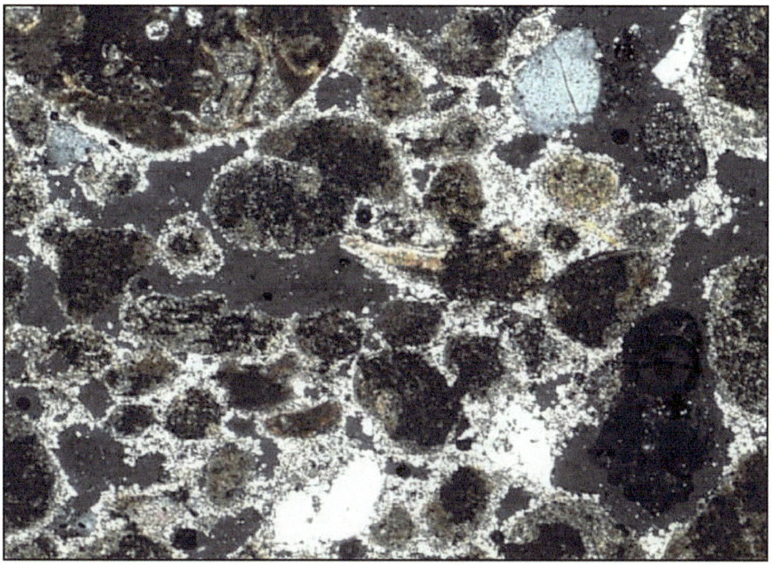

(a)

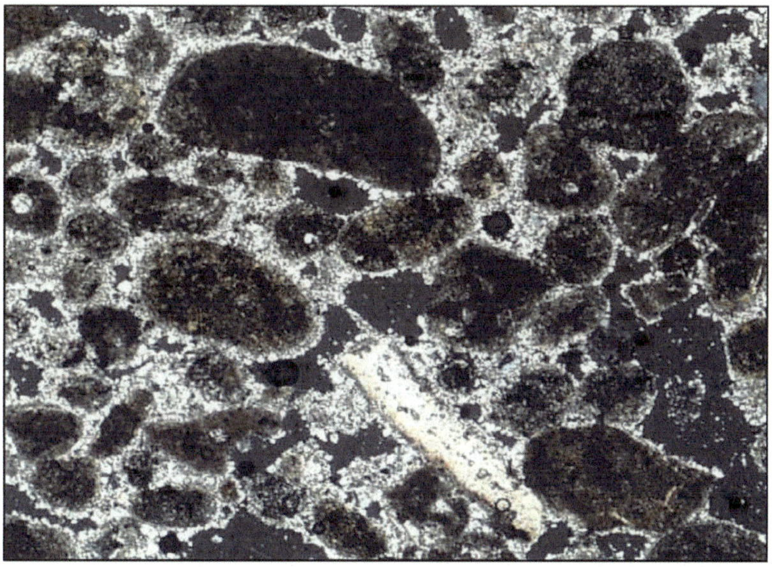

(b)

Figure 4.
(a, b) Photomicrograph of intact calcarenite under cross polarized light (XPL) showing wackestone (pele-oo-sparite) texture with drusy sparite, Mustafa Kamel Necropolis (Weathered sample, heterogeneous pore system).

enables the assessment of pore size and distribution in relation to the distribution and formation of the minerals involved.

4.4 Comparison between the sound and weathered rock layers

Samples can be clearly distinguished from the alveolar portions - the amortized and non-woven parts using thin, unpainted limestone sections that feature a relatively homogeneous pore structure. In contrast to unpainted areas, alveolar flats have a heterogeneous pore structure, for example pores often contain ferric oxides and hydroxides indicating a lower total pore size and higher content of small spots.

(a)

(b)

Figure 5.
(a, b) Photomicrograph of fossiliferous oolitic intraclastic limestone thin section under cross-polarized light (XPL) showing subrounded monocrystalline quartz grains (QTZ) and porous region, El-Shatby Necropolis.

Data from microscopic polarization and electron microscopy experiments show that oxygen clarity of NaCl crystals is strongly influenced by the rate and volume of moisture changes, and how they shrink with changes in crystal size.

5. Creep tests (materials and experimental program)

Creep is an irreversible ductile deformation in time under constant stress. Creep strain seldom can be recovered fully when the loads are removed, thus it is largely "plastic deformation." It is a progressive phenomenon initiated at a certain time after excavation at a certain location around the profile and spreading in time into the rock mass. For the long-duration design life of underground structures, the long-term stability of the tunnel must receive major consideration. For this reason, time-dependent deformation behavior of the surrounding rock must be well understood. Neglecting creep effects during deep excavation may lead to incorrect evaluation of deformation and thus may impact on the criteria for selection of proper design.

Understanding the mechanisms of rock breakdown that have been excavated within ancient monuments requires a thorough study of the mechanical behavior of these rocks, and the importance of the physical and mechanical properties of these rocks to understand the phenomena of instability.

The results of the geotechnical characterization of these rocks will be used in numerical modeling and design of reinforcement measures. For this purpose, a new laboratory testing program will be launched.

Rocks, sample preparation, experimental setup used and the procedure are briefly described below.

5.1 Types of creep

An idealized creep curve for rock at constant stress consists of three stages: instantaneous elastic strain followed by primary creep with decreasing creep rate, then steady-state creep with constant creep rate, and finally tertiary creep with increasing creep rate leading to failure. Most of the work on time-dependent strain has been conducted on primary and secondary creep phases only and the tertiary phase has not been investigated in appreciable detail.

5.2 Laboratory test specimens

In this study, the size of the comprehensive laboratory testing program using cylindrical samples with 42-44 mm diameter and height (91–103 mm). Although these rocks do not show distinct layers, the nuclei were extracted from the blocks and their masses in the vertical direction, which was expected to represent the physical properties of these units perpendicular to the layers. However, some samples were also extracted in a vertical direction on the mattress. Some specimens were broken and/or small cracks or cracks appeared on their surfaces. However, in order to achieve reliable assessments, the number of samples was increased as many as possible. Laboratory tests were performed in accordance with the testing procedures proposed ISRM and recommended by ASTM at the Engineering Geology Laboratory, Department of Civil Engineering, University of Aristotle Thessaloniki, Greece.

5.3 Laboratory tests

Laboratory studies (experimental examination) were performed on surface rock samples and prepared surfaces. The basic mechanical testing of the laboratory includes the behavior of deformation to failure under uniaxial and triaxial compression and we offer a complete creeping rock characterization conducted during the past 2 years from a series of isotropic and isotropic compression tests conducted in the inventory of various stresses, viscosity behavior was determined by following a procedure, the multi-step download, which emphasizes the transit creep side.

5.4 Very slow uniaxial creep tests on calcarenitic and sandy oolitic rock specimens under investigation

Creep in hard brittle rocks is rare as deformation rate is extremely slow. Hard rock shows creep behavior appreciably only at elevated temperatures and pressures generally not encountered in engineering structures. Soft rocks on the other hand creep mostly at the room temperature, atmospheric pressure, and deviatoric stress range normally encountered in engineering structures.

5.5 Analysis of creep behavior of soft rocks in tunneling

Regarding viscous plasticity, despite much work done on high porous rocks, only over the past years, there has been growing concern about the long-term behavior of deep underground structures in general. The rock mass tests large strain rates of viscosity and plastic. However, after a few years, the stress rates become smaller and reach a fairly stable condition characterized by very small stress rates.

It is known that most rocks have time-dependent behavior, and the viscous and plastic modeling of rocks and soils is of great importance both in petroleum engineering and underground engineering, for example when assessing deformations at the walls of deep fossil sections or considering pressure problems.

Moreover, when smaller time periods are considered, the stress distribution around a cave or exposures is such that the divergent pressure decreases rapidly with respect to the distance to the cave. Very small stress rates are tested at large distances within the rock mass and should be evaluated when predicting the behavior of the cave or photo gallery [40].

The limited available literature may be rooted in the particular problems raised by the long-term creep test, in the short term, as described below.

1. When the creep rate is $10 = 10^{-12}\ s^{-1}$, a 12-day test results in a strain of $\varepsilon = 10^{-6}$. The coefficient of thermal expansion of rocks is in order $\alpha = 1^{-4} \times 10^{-5}\ C^{-1}$, that is, the "noise" (i.e., elastic thermal deformation sample) due to small temperature changes will be greater, in most cases, than the signal to be measured (e.g., the average sample deformation arose from proper creep). The same can be said for moisture variations, which have a significant impact on many rocks.

2. Slow creep rates are obtained when small mechanical loads are applied. Most of the crawl test devices are designed to work in a DVR pressure range of 5–20 MPa. Stress control is usually weak when the applied pressure is less than 1 MPa.

3. The creep rate is calculated by comparing strains $\varepsilon_{1,}$ measured in two different times, τ_1 and $\tau_{2,}$ or $\varepsilon. = (\varepsilon_2 - \varepsilon_1) / (\tau_2 - \tau_1)$. When the compression rate is in the range $\varepsilon = 10^{-12}\ s^{-1}$, it can be reasonably evaluated on a daily basis ($t_2 - t_1 = 105$ s, $\varepsilon_2 - \varepsilon_1 = 10^{-7}$) only if ε_1 and ε_2 can be measured with an accuracy of not less than 10.8, or one-tenth of the expected difference between the two successive measured breeds [40].

Tightening of fragile rocks results in distributed damage long before the rocks fail unstable. The damage is usually manifested in small fractures and expansive microcracks [41–43]. These small fractions are usually smaller than the grain size and are often distributed almost uniformly before they are locally cracked. There are no uniform distributions of small fractions associated with the nucleus of error and growth.

Partial damage was used to explain the reduction of seismic wave velocity, earthquake variation, reduction of elasticity and strength units, and rock failure mechanics. In addition, stress damage can facilitate time-based creep-driven by stress erosion and subcritical crack growth. This creep strongly affects long-term strength and failure stability. For example, granite samples that are exposed to 1 month of non-axial static pressure under a pressure of approximately 0.65 may fail—or "delayed fractures" may develop days to years after removal of applicable loads.

The creep test shows how strain builds up over time under constant pressure. The rock usually deforms quickly and then begins to deform more slowly after the yield fatigue, which is called the initial creep. After the initial creep (I), the deformation continues at a constant rate in the linear part of the curve, which is secondary creep (II). Finally, the deformation rate increases rapidly until the rock fails to "fracture" in the high creep (III), if stress is removed but the strain remains permanent.

Three stages of creep behavior can be identified: in the first stage, they are classified as initial creep, and strain occurs at a decreasing rate. In some cases, the primary creep curve approaches a constant rate of strain called secondary creep. In high-stress specimens, secondary creep may turn up in higher creep, which is characterized by an increased strain rate until creep failure occurs suddenly. In the last two stages, the thin vertical cracking begins, accompanied by hardening, and only near failure, large cracks spread rapidly and lead to a sudden collapse. Long-term tests performed on a secondary creep sample revealed even appearance at 40% of estimated strength. The purpose of this research is to make recommendations on the promotion and safety of long-term underground historical structures under load. For this purpose, there is a set of experimental tests and advanced numerical analyses.

6. Description and discussion of the experimental program

6.1 Describe full creep tests

The research demonstrates an integrated empirical approach aimed at assessing safety and strengthening historic underground structures under high pressure.

The purpose of these tests is to obtain data, first, to determine the amount of sticky parameters that govern the long-term behavior of these structures, and secondly, to validate numerical models.

6.2 Testing device

Long-range uniaxial creep tests were performed on standard cylindrical rock samples collected from the three archeological sites under investigation (diameter D = 4.2–4.4 mm, height H = 90–103 mm); samples were prepared for testing according to ASTM standards with length-to-diameter ratios approximately 2.25, all samples have highly polished end surfaces to minimize final effects. The sample was set between two solid steel plates, with a steel cover between the sample and the two plates. During each test, two high-precision displacement sensors at two vertical levels at a 90° angle allowed both the relative rotation of the two pages and the measurement of the average relative displacement.

6.3 Sensors

Applied loads and the resulting strain were recorded using an automatic data acquisition system, sampling at a rate between 1 and 3 readings per second, thereby overcoming any deficiencies in data resolution.

6.4 Loading

The approved test procedure consisted of loading samples at a constant rate of about 1.35 MPa up to 1.75 MPa for samples from Catacomb in Kom El-Shoqafa, 1.55 MPa up to 2.17 MPa for samples from Mustafa Kamel Necropolis, and 2.6 MPa up to 3.44 MPa for samples from El-Shatby cemetery. In order to keep the applied pressure as stable as possible, dead weights were used and steel cylinders were placed on the upper steel plate on the upper face of the cylindrical sample. The applied stress is calculated by dividing the weight of the steel cylinders placed on the top plate by the initial cross-sectional area of the sample.

6.5 Temperature and hygrometry

The temperature changes during a long-term creep test must be as a small as possible and must be measured precisely enough to allow correction of the raw strain data for thermoelastic strains; in our study, all the periods of test were in the room temperature between 24 and 26° in the laboratory by controlling the air condition.

7. Test results

Uniaxial creep tests were performed on three rock samples from each site. Rock samples are loaded through fixed uniaxial compression at 1, 35, 1, and 75 MPa (one stress per sample) for Catacomb of Kom El-Shoqafa rock samples collected, at 2.6 and 3.44 MPa for rock samples collected from El-Shatby archeological site, and at 1.55 and 2.17 MPa for rock samples collected from Mustafa Kamel Necropolis.

The experimental procedure follows ASTM standards (ASTM D4405 and D4341). The compression machine is used to apply continuous axial load to the samples. Digital scales are installed at 0.001 millimeters to measure the axial displacement of the samples, see **Figures 6–9**. Samples are loaded continuously for 1 to 2 years until the samples fail without any acceleration, depending on the displacement results. During testing, axial distortion, time, and failure modes are recorded. The readings are repeated every minute at the beginning of the test, and gradually decrease to twice a day after the first few days of testing. This also depends on the deformation rate of each sample. The results are presented by strain time curves. Axial stress and axial pressure values are calculated by:

$$\sigma_{axial} = Pa/A, \tag{1}$$

$$\varepsilon_{axial} = \Delta L/L, \tag{2}$$

where σ axial is the axial pressure, Pa is applied axial load, A is the normal cross-section area of the direction of the load, ε_{axial} is the geometric axial strain, ΔL is the axial deformation, and L is the original length.

Table 2 summarizes the results of a uniaxial creep test. The axial stress time curves are shown in **Figures 10–16**, and the curves represent instantaneous, transient, and triple creeps of rock samples under a fixed axial load. Samples are loaded quickly and then the axial strains increase. The immediate breeds range from 0.07 to 3.5.

Most samples, under constant axial pressure, show a complete creep stage: transient, steady, and triple creep stages.

Increasing the value of the instantaneous creep strain with hard axial stress gives strain time curves of rock samples tested under constant high and low axial

Figure 6.
Rock creep testing devices. Samples are 90–105 mm high, 42–44 mm² diameters. Two displacement sensors were used during each test.

Figure 7.
The collected intact sandy oolitic limestone specimens from El-Shatby Necropolis site under creep testing devices.

pressures. Axial stress also increases crawling strains. In the transit crawl stage, the stress rate increases with the applied stressors. In most cases, the stress rate under high axial pressure is greater than the low axial pressure rate. The effect of embedding in the sample may make the compression rate under low pressure higher than the pressure under high pressure.

7.1 Catacomb of Kom El-Shoqafa, Test No. 1

On May 5, 2016 (Day 1), Catacomb of Kom El-Shoqafa no_1 began testing on a sample of sandy limestone, loading it to a vertical stress of $\sigma 1$ = 1.75 MPa, 65% of the coaxial compression strength of the rock material (peak sample strength). **Figure 10** displays the strain curve versus time; this curve averages the data provided by two displacement sensors. Strains do not correct for elastic thermal differences. In this test, the crawl was faster than the Catacomb of Kom El-Shoqafa site. Test no_2: From day 130 to day 200 after the start of each test, the cumulative strain was 4.5

Figure 8.
The collected sandy oolitic limestone specimens from the catacombs of Kom El-Shoqafa site under creep testing devices.

Figure 9.
The collected intact calcarenitic rock specimens from Mustafa Kamel Necropolis site under creep testing devices.

microns for Catacomb of Kom El-Shoqafa. 1 and 2.8 microns for Catacomb of Kom El-Shoqafa test site 2. This difference is fully in line with what is known in previous tests conducted at greater pressures on these samples. When the stress rate in the transient pressure zone is increased, followed by a similar decrease, it can be observed from day 44 to day 130, immediately followed by a steady slope (steady state crawl) up to 205 days. Finally, a more stable condition followed with a smaller stress rate until the sudden sample failure on day 368. Stress rate developments were more progressive in this case. There is no specific explanation for these changes in compression rate. At the end of the test, the observed pressure rate is $\varepsilon = 2.30 \times 10^{-8}\ s^{-1}$, the sample was suddenly broken after the 368 day (end of the test) on June 1, 2017, while the sample showed a higher creep phase.

7.2 Catacomb of Kom El-Shoqafa, Test No. 2

On September 1, 2016, (Day 1) after the start of the previous tests, an identical creep device on the same table was assigned to the catacomb of Kom El-Shoqafa

Specimen No.	Testing period	Time (days)	
		1 9 100 135 178 375 667 786 813	
Catacomb of Kom El-Shoqafa test N_1 (sandy oolitic limestone)	From **5/5/2016** to **1/6/2017**	σ_1 = 1.35 MPa	
Catacomb of Kom El-Shoqafa test N_2	From **1/9/2016** to **2/7/2018**	σ_1 = 1.75 MPa	
El-Shatby Necropolis. test N_1 (oolitic intraclastic limestone)	From **12/4/2016** to **4/7/2018**	σ_1 = 2.60 MPa	
El-Shatby Necropolis. test N_2 (oolitic intraclastic limestone)	From **5/5/2016** to **3/7/2018**	σ_1 = 3.44 MPa	
Mustafa Kamel Necropolis .test N_1 (intact Calcarenite)	From **3/4/2016** to **22/3/2017**	σ_1 = 1.55 MPa	σ_1 = 1.86 MPa
Mustafa Kamel Necropolis. test N_2 (intact Calcarenite)	From **11/4/2016** to **28/3/2017**	σ_1 = 2 MPa	
Mustafa Kamel Necropolis. test N_3 (intact Calcarenite)	From **6/4/2016** to **7/4/2016**	σ_1 = 2.50 MPa	

Table 2.
Uniaxial creep test, testing program.

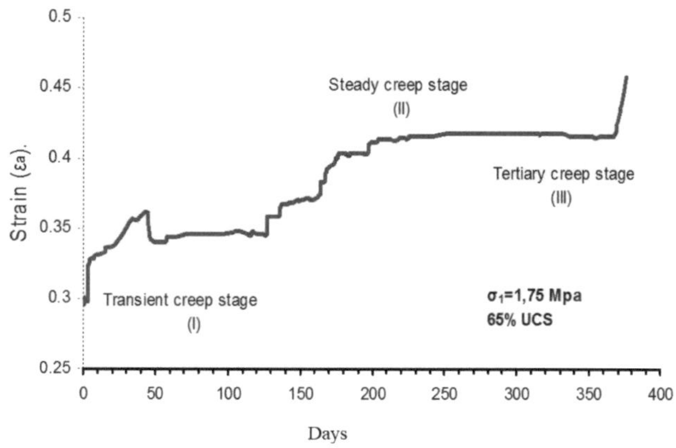

Figure 10.
Strain versus time curve during the catacomb of Kom El-Shoqafa, uniaxial creep test no. 1.

Test no_2, which began on another sample significantly purer than the previous, loaded on a vertical stress σ1 = 1.35 Mpa, 50% of the axial compression strength of the rock material (peak strength), the applied stress until the end of the test was not adjusted without sample failure on July 2, 2018, during a steady slope or steady state and a creep with a small strain rate was observed. **Figure 11** displays a curve versus time. This curve averages the data provided by two displacement sensors. The compression rate (ε.) is calculated every 5 days; it is calculated for 10 days. Strains are corrected for temperature variation. Initially, the strain experienced a long initial transient period until the first few days characterized by a slow decline in rate, with the average stress rate stabilizing to $\varepsilon = 5.85 \times 10^{-10}\ s^{-1}$ (positive sample contractions), with long-term amplitude fluctuations ++20%; this is probably associated with moisture fluctuations. This phase was followed by a long steady slope or steady-state creep to the end of the test while the observed compression rate was ε. $= 3.21 \times 10^{-9}\ s^{-1}$, while it was $1.50 \times 10^{-9}\ s^{-1}$ was at the beginning of the test.

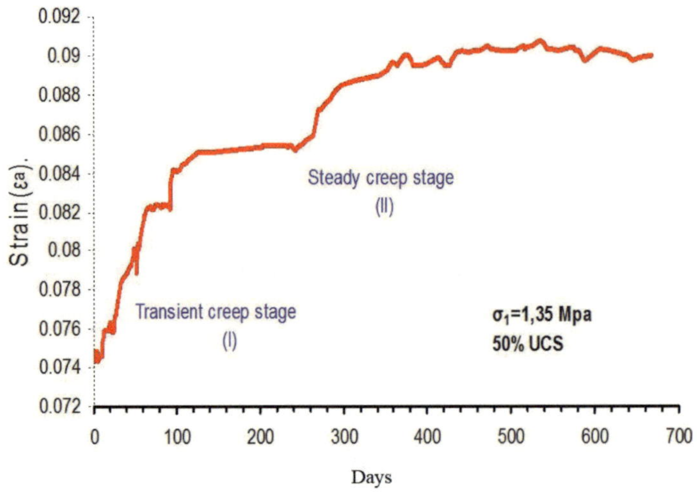

Figure 11.
Strain versus time curve during the catacomb of Kom El-Shoqafa, uniaxial creep test no. 2.

7.3 El-Shatby Necropolis, Test No. 1

On April 12, 2016, (Day 1) testing of Shatby Tombs No. 1 began on a sample of sound rock-limestone that was loaded to = 1 = 2.60 MPa, 50% of the axial compression strength of the rock material (peak strength) is not Stress adjustment until the end of the test on July 4, 2018. **Figure 12** shows the stress curve versus time, where the elastic strain is followed by a long transient creep characterized by a slow rate of decline, followed by a slope or creep constant in a steady state with a small stress rate until end of the test without sample failure; this curve averages the data provided by two displacement sensors. The compression rate (ε.) is calculated every 1 h at the beginning of the test, and after the first few days it is calculated every day. Strains are corrected for temperature variation. The strain experienced a long initial transient period, where the average stress rate stabilized on $\varepsilon = 1.3 \times 10^{-9}\ s^{-1}$ (positive sample contractions.), with long-term amplitude fluctuations ++15%; this is probably associated with moisture fluctuations.

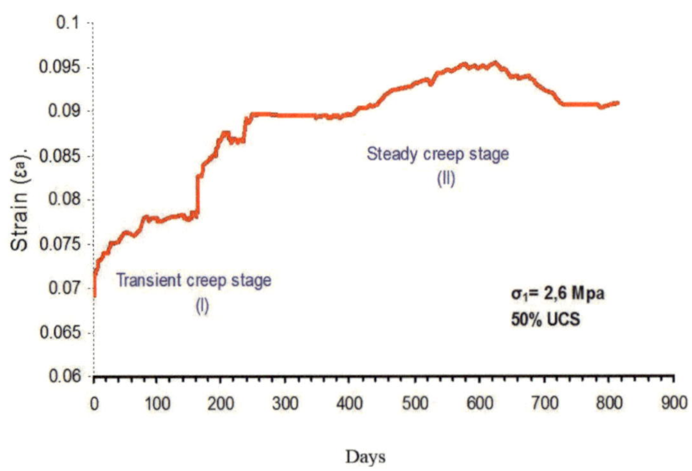

Figure 12.
Strain versus time during El-Shatby Necropolis, uniaxial creep test no. 1.

Transient reverse crawl was observed on day 214 to day 244, sometimes referred to as "hypotension." During this test, this reverse crawl lasted much longer (20 days) than is currently observed in tests with greater stress. The stress rate stabilized one way or another after day 260, but at the end of the test, the observed pressure rate was $\varepsilon. = 1.62 \times 10^{-9}\ s^{-1}$.

7.4 El-Shatby Necropolis, Test No. 2

On May 5, 2016 (after the start of the previous tests), an identical crawl device was assigned to the same table, and El-Shatby test of Q2 was started on a cylindrical sample with geometric dimensions similar to that used in El-Shatby test of cemetery no_1, loaded on a vertical stress of $\sigma 1 = 3.44$ MPa, 65% of the coaxial compression strength of the rock material (**Figure 13**). A long transient period can be observed followed by a constant inclination or a steady-state crawl until the last day of recording. In this test, the crawl was faster than at El-Shatby Cemetery, test number 1: from day 70 to 270 after the beginning of each test, the cumulative strain was 2.5 μm for El-Shatby Cemetery, test number 2 and 1.8 microns for El-Shatby Cemetery site, test number 1. This difference corresponds exactly to what is known from previous tests conducted at greater pressures on these samples. An increase in the stress rate can be observed, followed by an equivalent decrease, at day 260 and at around day 324, and stress rate developments were more progressive in this case. There is no specific explanation for these changes in compression rate. At the end of the test on July 3, 2018, the observed pressure rate was $\varepsilon. = 3.41 \times 10^{-10}\ s^{-1}$.

7.5 Necropolis of Mustafa Kamel, Test No. 1

On April 3, 2016, (Day 1) Mustafa Kamel's # 1 test began on an intact sample initially loaded at 1.55 MPa but no creep was observed until 9 days after the test began. Perhaps the pregnancy is too small to produce any detectable strain. Thereafter, the applied pressure was adjusted once, and was constructed up to $\sigma 1 = 1.86$ MPa (+10%) after the 9th day 60% of the uniaxial compression strength of the rock material. The numbers in parentheses indicate that the compression value is adjusted. **Figure 14** shows the pressure curve versus time; this curve averages the data provided by two displacement sensors. The compression rate ($\varepsilon.$) is calculated

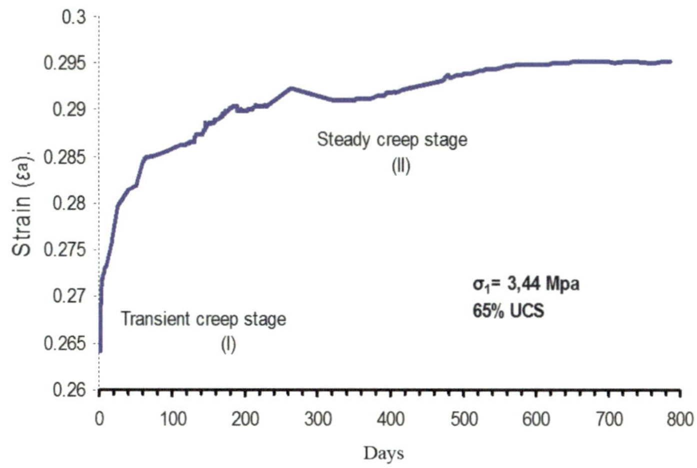

Figure 13.
Strain versus time curve during El-Shatby Necropolis, uniaxial creep test no. 2.

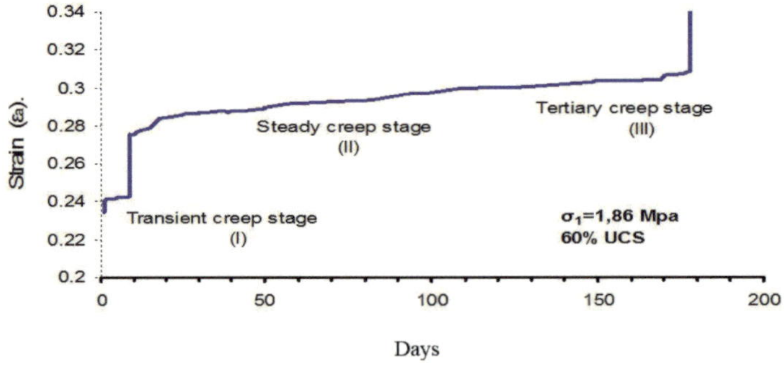

Figure 14.
Strain versus time curve during Mustafa Kamel Necropolis, uniaxial creep test no. 1.

every 5 days; it is calculated for 10 days. Strains are corrected for temperature variation. The strain experienced a long initial transient period characterized by a low slow rate followed by a steady slope or a steady-state creep with a small stress rate, at which time the average stress rate stabilized to $\varepsilon = 1.62 \times 10^{-9}\ s^{-1}$ (positive sample contractions.), with long-term capacity fluctuations of + _20%; this is probably associated with moisture fluctuations.

The transient inverse creep has not been observed, and is sometimes referred to as "stress drop." Strain rate more-or-less stabilized after day 160, and strain rate $\varepsilon^{\cdot}$ = $4.86 \times 10^{-9}\ s^{-1}$ and the sample has been broken suddenly after 178 days (the end of the test 22/3/2017); the specimen showed the complete three phases of creep end with the tertiary or acceleration creep stage.

7.6 Necropolis of Mustafa Kamel, Test No. 2

On April 11, 2016, (after the start of the previous tests) an identical crawl device was set on the same table, and Mustafa KAM # 2 test started on another sample that is significantly purer than the previous, loaded on the stress of σ1 = 2 MPa, 65% of the uniaxial compression force for rocky materials, and applied pressure was not modified until the end of the test on 28/3/2017 (**Figure 15**). It displays the strain curve versus time; this curve averages the data provided by two displacement sensors. In this test, the creep was faster than the site of Mustafa Kamel's tombs, test number 1: from day 11 to 91 after the start of each test, the cumulative strain was 3.7 microns for the Mustafa Kamel test site Necropolis. 2 and 3.5 microns of the graves of Mustafa Kamel site No. 1. This difference corresponds exactly to what is known from previous tests conducted at greater pressures on these samples. An increase in

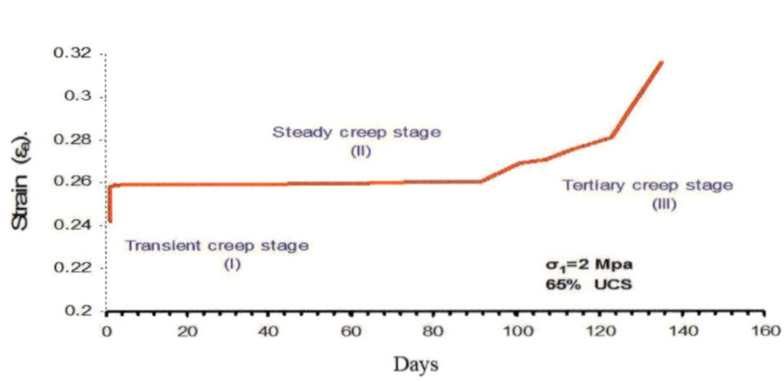

Figure 15.
Strain versus time curve during Mustafa Kamel Necropolis, uniaxial creep test no. 2.

stress rate was not observed in this test, followed by an equivalent decrease, while a long transient strain was encountered and a slow decline in rates was followed by a creeping phase in a steady state with a very small stress rate until day 91, after acceleration or the third stage of creep began. The 135th day in a large stress rate $\varepsilon. = 1.11 \times 10^{-9}\ s^{-1}$.

7.7 Necropolis of Mustafa Kamel, Test No. 3

On April 6, 2016, an identical creep device was set on the same table, and Mustafa KAM # 3 test was started on another heavily loaded sample on a stress of $\sigma 1 = 2.5$ MPa, 80% of the uniaxial pressure force of the material rock (peak strength). Strains do not correct for elastic thermal differences. **Figure 16** displays a curve versus time. This curve averages the data provided by two displacement sensors. In this test, the crawl was faster than the site of Mustafa Kamel's tombs, test number 1 and test number: from the first day after the start of the test, the cumulative strain was 7 microns for the site of Mustafa Kamel Necropolis, test number 3. This difference corresponds exactly to what is known from the tests previously conducted at smaller pressures on these samples. The sample fractured 26 h after the start of the test, the crawl begins with a short elastic strain followed by a short transient strain followed by a steady-state crawl with a very small stress rate up to 23 h after the start of acceleration or triple crawl resulting in a sudden failure of the sample with a high stress rate after 26 h exactly. At the end of the test, the observed pressure rate was $\varepsilon. = 0.30$ per second.

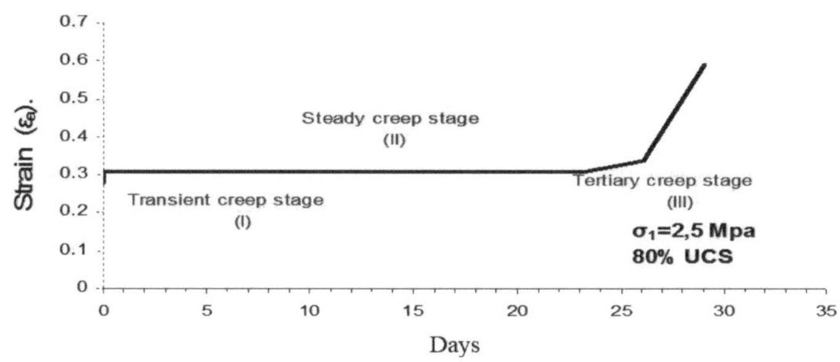

Figure 16.
Strain versus time curve during Mustafa Kamel Necropolis, uniaxial creep test no. 3.

7.8 Comparison with tests performed under larger stresses

Qualitatively, the behavior of soft rocks under small pressure (0.1 = 0.1–3 MPa) exhibits the same general features as observed under large pressures (e.g., σ = 5–20 MPa). The rapid accumulation of stress leads to a transient creep characterized by a slow rate of decline. The creep rate then becomes almost constant (a steady state is reached) or, more precisely, its average value remains constant, but the rate faces long-term fluctuations that may be affected by slow changes in moisture measurement. Reducing the load ("low pressure") creates an inverse crawl, which lasts much longer during tests under greater stress.

Norton-Hoff's constitutive equation is often proposed to describe stable state creep.

$$\varepsilon^{\cdot} = A^{*}\ \exp.(-Q/RT)\ \sigma^{n} \tag{3}$$

where σ is the applied deviatoric stress; T is the absolute temperature; and A*, n, and Q/R are constants. For Etrez salt, Pouya suggests the following parameter values:

$$A^* = 0.64\ \mathrm{MPa}^{-n}\ \mathrm{yr.}^{-1}, Q/R = 4100\ \mathrm{K}, \text{and}\ n = 3.1. \quad (4)$$

Berest et al. [40] found that if the Norton-Hoff Law of Conditions was derived in Creep Test 1 (σ = 0.108 MPa, T = 286.5 K), the calculated compression rate ($\varepsilon.$ = $10^{-17}\ s^{-1}$) is smaller. Start by from the observed compression rate ($\varepsilon.$ = $1.4 \times 10^{-12}\ s^{-1}$). The observed pressure rates, even if they are too small, are much larger than expected. Spears et al. suggest that the pressure solution (rather than infiltration and slip, the mechanism that controls high stresses) is the most effective mechanism for crawling at very small pressures; the exponent of stress in this context would be n = 1 instead of n = 3–5, which is observed during standard tests. If this proposal is adopted, the Creep law should be modified when considering small pressures, with significant consequences in predicting the cave or gallery convergence rate.

Many lessons were learned during the test under these unusually low pressures. This first series of tests opened the way for further research on the behavior of rocks under very small pressures, long-term single-axis crawl tests were performed for geological and engineering applications on rock samples (for 850 days), and the applied loads were as small as 1.35 MPa. Slow stress rates such as $1.11 \times 10^{-10}\ s^{-1}$ were observed in some cases. These small loads and pressure rates pose several specific problems: potential drift of sensors during long 2-year tests, interference with small changes in room temperature and moisture measurement, and effects related to irregular load distribution applied to sample surfaces. These difficulties have been recognized and at least partially addressed. The qualitative results are in good agreement with what is known as the behavior of soft rocks under greater pressure; however, the observed pressure rates, even if they are extremely small, were much greater than expected.

The initiation, accumulation, and growth of cracks caused by stress in rocks are generally referred to as rock damage. Referring to the pressure caused by the crack is the load at which the sample will eventually fail, under prolonged loading, which they propose correspond to about 70–80% of the peak strength of the sample. It is also believed that the damage to the crack damage or the crack damage threshold point corresponds to the point at which the stress reflection or sample expansion begins. Corresponding to the volumetric stress gradient is approximately 70% of the estimated unrestricted compressive strength of the rock.

These stresses are well above the stress threshold for damage. It has been suggested that sample composition for unrestricted compression force tests reduces the spread of cracks. Many researchers suggest that the strain of the ring is generated between a stretch crack and the outer surface of a cylindrical sample. This breed may generate a confinement collar that limits the growth of continuous cracks.

Preliminary test results suggest that an alternative mechanism may affect the spread of unstable cracks. Under pure uniaxial loading conditions, a split can be expected parallel to the maximum pressure direction. The failure may ultimately be at the microscopic level due to the curvature of the rock slabs resulting from tensile fractures directed toward the maximum compressive pressure, as shown in **Figure 17**.

7.9 Triaxial creep tests on the calcarenitic and sandy oolitic rock specimens under investigation

The purpose of triaxial creep tests is to determine the viscosity and plastic parameters of the soft rock samples under confined conditions and to investigate

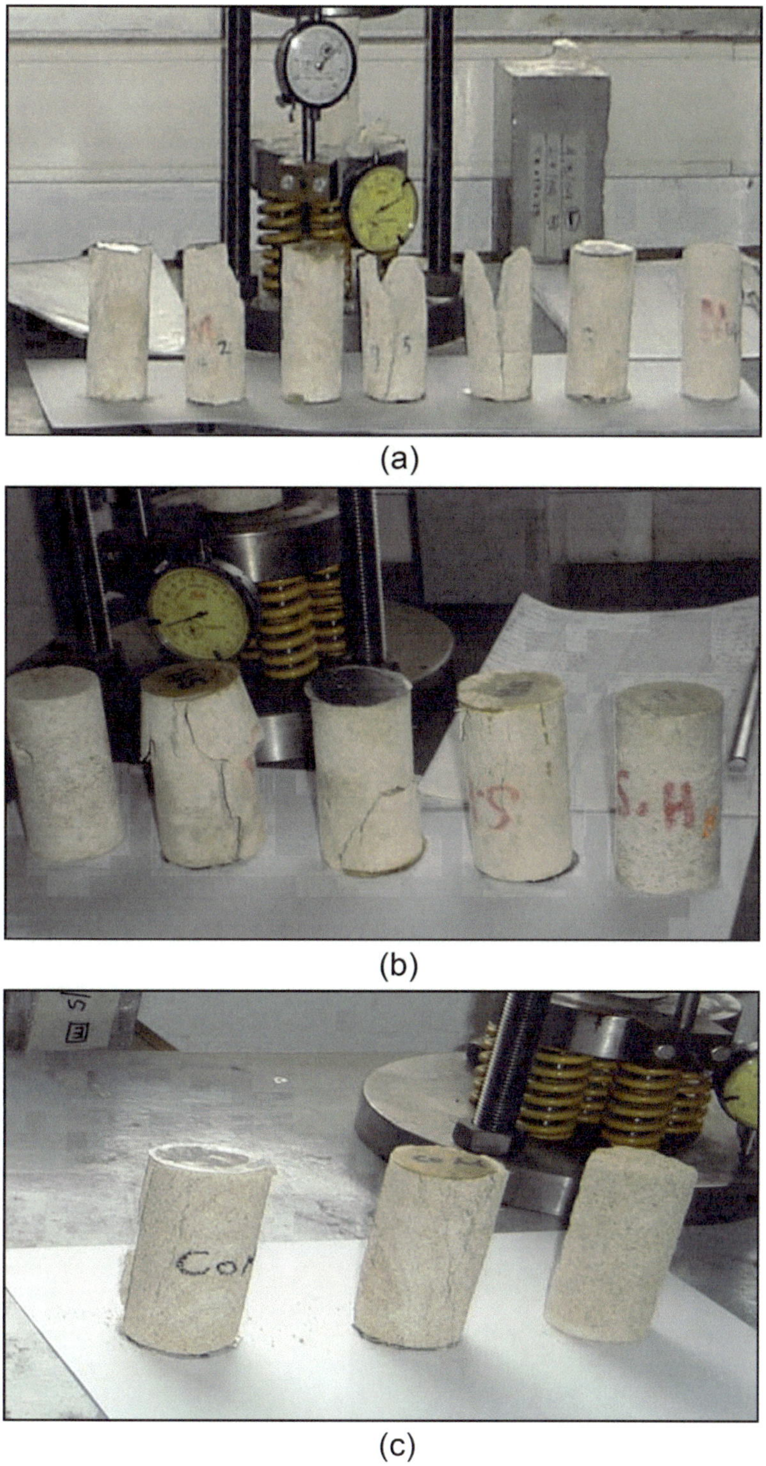

Figure 17.
Rock specimens under investigation, after uniaxial creep test. (a) Calcarenitic rock specimens, Necropolis of Mustafa Kamel. (b) Oolitic intraclastic limestone specimens, El-Shatby Necropolis. (c) Sandy oolitic limestone specimens, Catacombs of Kom El-Shoqafa.

the effects of axial stress and fortified pressure. Time-related parameters are monitored, recorded, and analyzed.

7.10 Test methods

Two samples of rock (length = 91–103 mm, diameter = 41–44 mm) were tested from each site under different constant axial pressures and different static pressure pressures for approximately 300 h. The experimental procedure follows the ASTM standard (ASTM D4406-93). The compression machine (fusion machine, 5000 kN) is used to apply the fixed axial load to the samples. Rock samples were placed in a three-axis cell (GDS) to provide constant confining pressure (**Figure 18**). The collected sample (Test # 1) of Catacomb of Kom El-Shoqafa is immediately loaded to the axial stress required at 1.45 MPa to limit the pressure by 225 kPa, and the applied axial stress was adjusted twice: the initial applied pressure was increased to 1 = 2, 17 MPa (+50%) after 98 h (2) then increased to $\sigma1$ = 2.53 MPa (+74%) after 125 h (3). The number in brackets refers to **Figure 18**, which displays the strain versus the time curve. Where the axial stress of up to 1.45 MPa and inventory pressures 510 kPa. Axial stress was not adjusted until the test ends after 200 h with steady-state creep with a small stress rate and without sample failure.

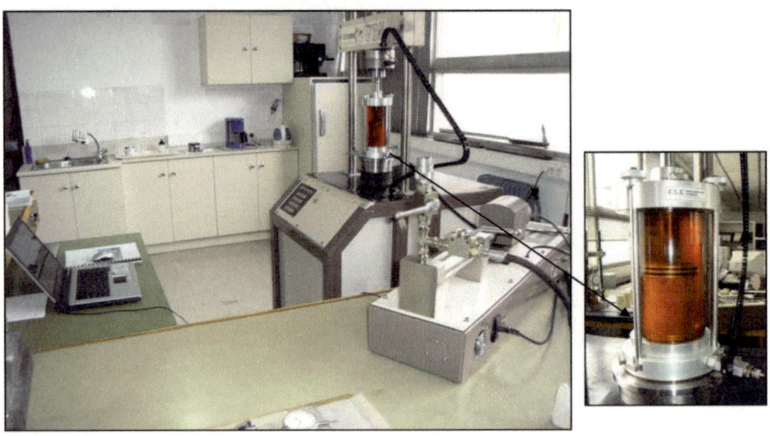

Figure 18.
Triaxial creep test device, with constant axial load under confining pressure. Triaxial creep test device. The cylindrical specimen placed inside (GDS) cell is loaded vertically using the compression machine.

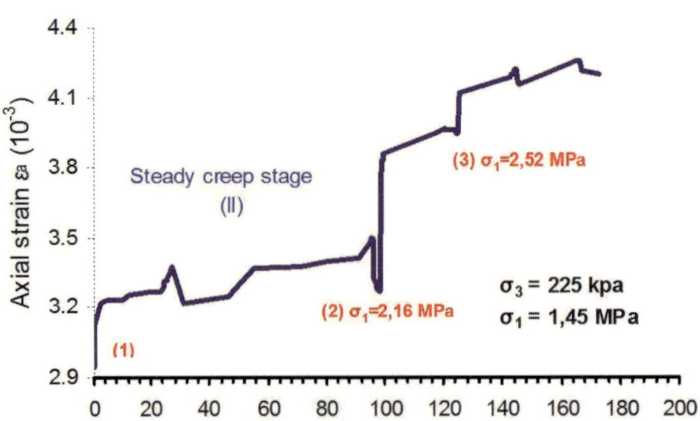

Figure 19.
Strain versus time during the catacomb of Kom El-Shoqafa, triaxial creep test no. 1.

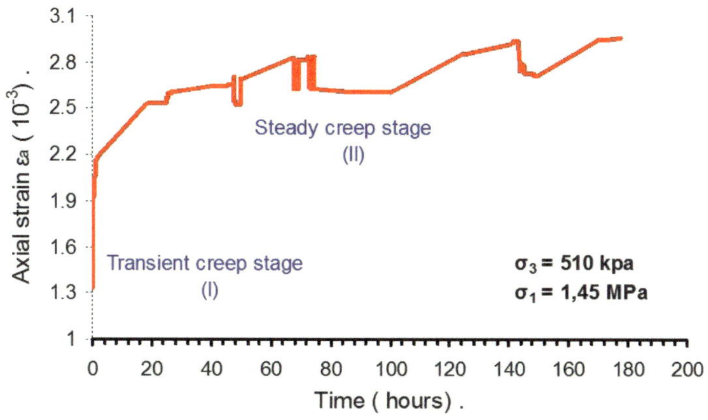

Figure 20.
Strain versus time during the catacomb of Kom El-Shoqafa, triaxial creep test no. 2.

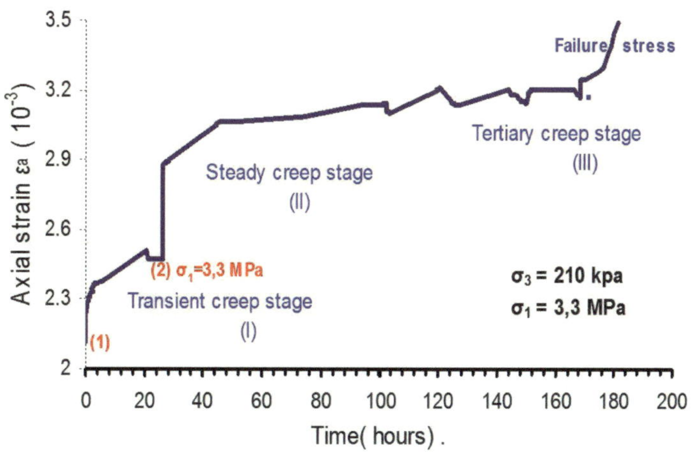

Figure 21.
Strain versus time curve during El-Shatby Necropolis, triaxial creep test no. 1.

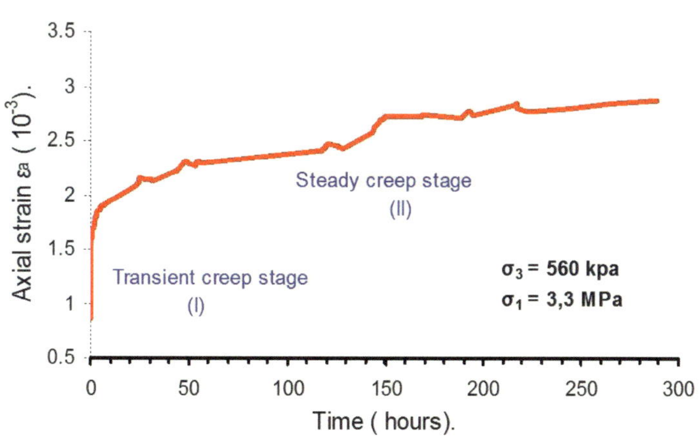

Figure 22.
Strain versus time during El-Shatby Necropolis, triaxial creep test no. 2.

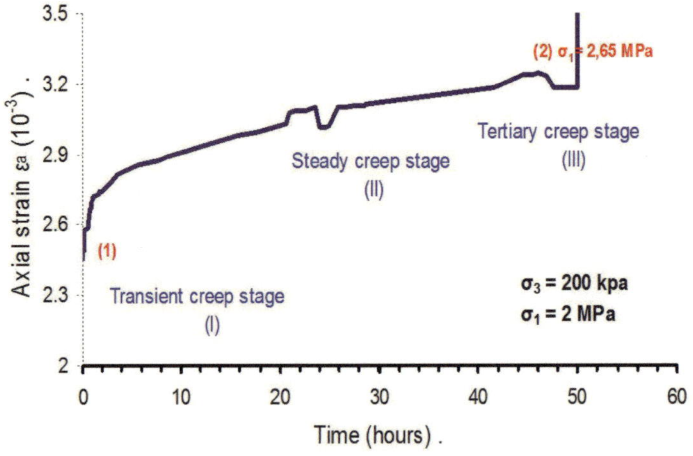

Figure 23.
Strain versus time during Mustafa Kamel Necropolis, triaxial creep test no. 1.

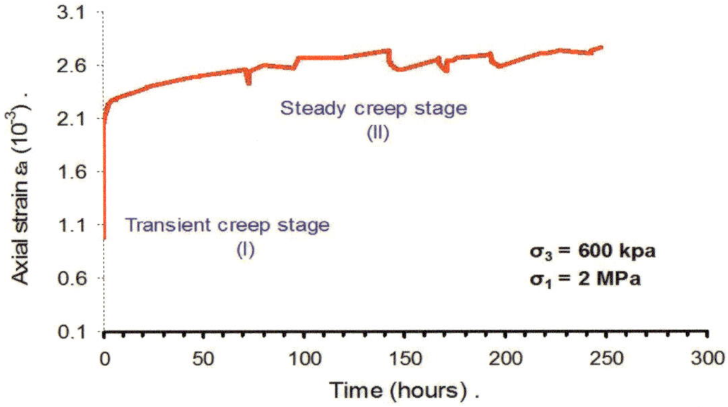

Figure 24.
Strain versus time curve during Mustafa Kamel Necropolis, triaxial creep test no. 2.

Samples collected (test # 1) from the Shatby Necropolis site were immediately loaded on the required axial stress at 2.63 MPa to limit pressure at 210 kPa, the applied axial stress was adjusted once: the initial applied pressure was increased to = 1 = 3.30 MPa (+25%) after 26 h (2).

In Mustafa Kamel Test No. 2, the sample was loaded on the axial stress required at 2 MPa to limit pressures of 600 kPa without modifying the axial stress until the end of the test at 300 h without high creep.

During testing, axial distortion and time are recorded. The frequency of reading is once every second at the beginning of the test, and gradually decreases to once every half an hour after the first day of the test. This also depends on the deformation rate of each sample. The results are presented by stress time curves in **Figures 19–24**. Axial stress and axial pressure values are calculated.

8. Test Results

8.1 Catacomb of Kom El-Shoqafa site

Axial strain time curves are shown in shapes (**Figures 19** and **20**). The curves represent transient and transient creep conditions of rock samples under constant axial load and compression pressure. Instantaneous strains were observed immediately after loading the range from 3×10^{-3} to 3.2×10^{-3} for the test number_1, and 1.3×10^{-3} to 2.2×10^{-3} for the test number_2. All samples show a long "slow low" primary transient creep and steady-state creep stages until the end of the test without acceleration or triple creep resulting in sudden failure. Observations on subsequent tests show that deformation increases rapidly at first to the first few hours of testing and tends to remain constant after that. Stress rates in a steady state are 0.01 to 0.02×10^{-3} h^{-1}.

In this test, it was observed that crawling during the catacomb of Kom El-Shoqafa site Test no_1 (where = 3 = 225 kPa) was faster than crawling during the catacomb of Kom El-Shoqafa site test no_2 (where = 3 = 510 kPa) with the same axial pressure σ1 = 1.45 MPa: From 1 to 96 h after the start of each test, the axial strain accumulated (10–3) 3.5 for the catacomb of Kom El-Shoqafa site Test No. 1 and 2.7 for the catacomb of Kom Shoqafa website Test No. 2.

8.2 El-Shatby Necropolis

Axial strain time curves are shown in shapes (**Figures 21** and **22**). The curves represent temporary and transient creeps of rock samples under constant axial load and confined pressure. Instantaneous strains were observed immediately after the loading range from 2.5×10^{-3} to 2.9×10^{-3} for number_1 test, and from 0.91×10^{-3} to 1.8×10^{-3} for number_2 test. All samples show a long "slow low" primary transient creep and steady creep stages (constant slope) up to the end of the test at 300 h except test number_1, which showed acceleration or triple creep stage leading to a sudden sample failure at 180 h where the confined pressure σ3 was small, that is, 210 KPa, while at test number_2, it was 560 kPa and the axial pressure was the same for the eyes σ1 = 3.3 Mpa. Observations on subsequent tests showed deformation increases rapidly at first to the first few hours of testing and tends to remain constant after that. The first sample failed after the end of the test. Pressure rates in the steady state are 0.01 to 0.015×10^{-3} h^{-1}.

It was observed that creep through Shatby site Necropolis, test number 1 (where k3 = 210 kPa) was faster than creep through Shatby cemetery site, test no_2 (where = 3 = 560 kPa) under the same axial pressure σ1 = 3.3 MPa: From 1 to 150 h after the start of each test, the accumulated axial strain (10^{-3}) was 3.2 for Shatby Necropolis site Test site 1 and 2.4 for Shatby site Necropolis Test No. 2.

8.3 Mustafa Kamel Necropolis

Table 3 summarizes the results of the triple axial crawl test. Axial strain time curves are shown in shapes (**Figures 23** and **24**). The curves represent transient and transient creep conditions of rock samples under constant axial load and compression pressure. Strains observed immediately after the loading range from 2.5×10^{-3} to 2.7×10^{-3} for test number_1, and from 0.98×10^{-3} to 2.3×10^{-3} for test number 2. All samples show a long initial transient creep "characterized by slow rate of decline" and steady creep phases (constant slope) up to the end of the test at 300 h except the first sample, which shows a transient, steady, and triple-accelerated creep phase leading to sudden sample failure at 49 h immediately after adjusting the axial pressure from σ1 = 2 MPa to σ1 = 2.65 MPa. Observations on subsequent tests showed deformation increases rapidly at first to the first few hours of testing and tends to remain constant after that. Pressure rates in the steady state are 0.01 to 0.015×10^{-3} h^{-1}.

In this test, it was observed that crawling through the site of Mustafa Kamel's cemetery in test 1 (where = 3 = 200 kPa) was faster than crawling through the site of Mustafa Kamel's cemetery. No_2 test (where = 3 = 600 kPa) under the same axial

Specimen No.	Testing period	Confining pressure (σ_3)	Time (h)		
			1 3 26 51 52 98 125 200 300		
Catacomb of Kom El-Shoqafa test N_1 (Sandy oolitic limestone)	From **19/10/2016** to **27/10/2016**	**225 KPa**	**σ_1 = 1.45 MPa**	**σ_1 = 2.17 MPa**	**σ_1 = 2.53 MPa**
Catacomb of Kom El-Shoqafa test N_2	From **21/11/2016** to **28/11/2016**	**510 KPa**	**σ_1 = 1.4 MPa**		
El-Shatby Necropolis test N_1 (oolitic intraclastic limestone)	From **9/11/2016** to **16/11/2016**	**210 KPa**	**σ_1 = 2.63 MPa**	**σ_1 = 3.30 MPa**	
El-Shatby Necropolis test N_2 (oolitic intraclastic limestone)	From **29/11/2016** to **11/12/2016**	**560 KPa**	**σ_1 = 3.31 MPa**		
Mustafa Kamel Necropolis test N_1 (Calcarenite rock)	From **16/10/2016** to **18/10/2016**	**200 KPa**	**σ_1 = 1.32 MPa**	**σ_1 = 1.98 MPa**	**σ_1 = 2.6 MPa**
Mustafa Kamel Necropolis test N_2 (Calcarenite rock)	From **12/12/2016** to **22/12/2016**	**600 KPa**	**σ_1 = 2 MPa**		

Table 3.
Triaxial creep test, testing program.

pressure = 1 = 2 MPa: from 1 to 45 h after the start of each test, the accumulated axial strain (10^{-3}) was 3.3 for Necropolis of Mustafa Kamel site Test no. 1 and 2.5 for the site of Mustafa Kamel cemetery test site 2.

Thus, the prevalence of cracking (in the fragile field) and pore breakdown (under high pressure conditions) are the prevailing deformation mechanisms of the selected rocks.

The cumulative results of various three-axis crawl tests, conducted at tight pressures ranging from 200 to 600 kPa, showed that crawling reduces the level of brittle stress on failure by 15–20% in relation to standard tests, and similarly, the resulting stress threshold (e.g., Pore breakdown (reduction)) is reduced by the same amount, while the volumetric component of the strain is diluted only in the absence of confined pressure, and shrinks completely even when $\sigma 3$ decreases.

The instantaneous creep strain depends on axial pressure and confining pressure. In general, increased continuous axial pressure leads to greater axial stress. The pressure rate under high axial pressure is greater than the pressure under the lower axial pressure for the same fixed pressure. The higher the confined pressure, the smaller the resulting pressure. Comparison of results obtained from other soft rocks/salts indicates that the stress rate depends on the stress and previous strain. This is also consistent with the conclusion of Courthouse and Ong et al. who describe soft rocks as close.

The time-based foundational model of soft rocks developed by Zhang et al. can reproduce the general crawl characteristics of soft rocks with high precision. The crawl failure time to load the strain of the aircraft is longer than that of the three-axis load because the strain load frame controls the sample to expand.

9. Micromechanics of creep in the calcarenitic rocks

There is now a large body of evidence that rock deformation at low temperatures and pressures occurs through two mechanisms widely referred to as faulty flow and ductile flow. The term ductile is often used in three different contexts, including (1) plastic deformation of single crystals, (2) homogeneous deformation or uniform flow, and (3) deformation over a certain amount of stress. Here we will use the term ductile in the macroscopic sense of homogeneous deformation where inferior microscopic processes include improved shear pressure, granulation, and granular flow.

Macroscopically, these microscopic processes form the flow of the calcite. Experimental evidence of Caracola flow includes (1) a broad shear area indicating distributed damage and intact granules in an extraction; (2) a large pore breakdown, often accompanied by small intracranial cracks caused by "fragmentation"; and (3) fractures. Unlike this distributed pervasive flow, the standard fragile deformation at low-effective pressures is characterized by an expansive fine fracture, leading to shear localization along narrower fracture zones, which often consist of sections linked to a zigzag pattern (e.g., [44–48]). In thin sections, the fragile fracture is evidenced by the presence of almost abundant small cracks, away from the shear fracture. Many of these miniature cracks are parallel to the main baseline pressure and may arise from axial splitting of healthy grains [42, 43] or cracking of grain boundaries.

From previous experimental studies, the researchers agreed that distributed sedimentary rocks, for calcarenitic sedimentary rocks, are the dominant failure mechanisms in highly porous rocks, especially at high effective medium pressures [49–52]. On the contrary, the fragile local fracture dominates the rocks with low porosity, as well as in high-porosity rocks with low effective pressure.

10. Conclusions

The catacombs of the Kom El-Shoqafa and Amod El-Sawari (Pompeii's pillar) site, located in the city center, 2.5 km from the sea coastline, are carved into the initial sandy limestone (cement limestone); Cross joints filled with fragmented sand and saturated with water in the lower parts. This unit is illustrated with loose sandstone. It is medium brown in color to decorate granulated limestone saturated with groundwater. It goes beyond the formation of the hayf (Pliocene) or the older myosin. Surface quadruple deposits obscure actual contact. The other two archeological sites, which are located close to the waterfront of Alexandria (Shatby Cemeteries, Mustafa Kamel Cemeteries), were excavated in internal limestone or calcite (coastal hills). Yellowish white upward become yellow brown bottom.

Based on tests carried out on air-dried samples prepared in the vertical direction, UCS values indicate that according to the classification adopted by the London Geological Society, which relies on the unrestricted compressive strength and the classification proposed by [33, 50]. These calcarenitic rocks from which excavations are carried out underground are classified as soft to very weak. It is also in good compliance with the Rock Quality Assignment System (RQD) for these types of soft rocks, where RR = 18 and RQD = 15–20% and a very poor quality range from 0 to 25. In addition, the results of static deformation tests indicate that the types of rock in question have high deformation.

It should be noted that the silica content at the Catacomb site in Kom El-Shoqafa is higher than in any area in Alexandria, possibly due to sedimentation processes, such as the high silica content that does not contain cement but is found as sand grains. In low rock durability and stiffness, high sand-like grain content reduces rock strength against salt crystallization and moisture pressures within rock pores. This is not only because of its high content of silica granules but also because it is a sparse rock. It is known that this type of limestone is characterized by low durability.

Three stages of crawling behavior can be identified by uniaxial and triple-axis crawling tests. In some cases, the primary creep curve approaches a constant rate of stress called secondary creep. In high-stress specimens, secondary crawl may turn up in higher creep, which is characterized by an increased stress rate until crawl failure occurs suddenly. In the last two stages, the thin vertical cracking begins, accompanied by hardening, and only near failure, large cracks spread rapidly and lead to a sudden collapse. Long-term tests were performed on a secondary creep sample showing even at 40% of estimated strength.

The weathering process is associated with structural properties, such as poor geotechnical properties, carbon chemical composition, the presence of soluble salts in the porous system, marine climate with characteristic humidity, and marine spray, groundwater.

Author details

Sayed Hemeda
Conservation Department, Faculty of Archaeology, Cairo University, Egypt

*Address all correspondence to: sayed.hemeda@cu.edu.eg

References

[1] Berest P, Charpentier JP, Blum PA. Creep of rock under small loading. In: Proceedings of SMRI Fall Meeting; Washington DC; 1999. pp. 11-24

[2] Costin LS. Time-dependent deformation and failure. In: Atkinson BK, editor. Fracture Mechanics of Rock. San Diego, CA: Academic; 1987. pp. 167-215

[3] Das S, Scholz CH. Theory of time-dependent rupture. Journal of Geophysical Research. 1981;**86**: 6039-6051

[4] De Meer A, Spiers CJ. Creep of wet gypsum aggregates under hydrostatic loading conditions. Tectonophysics. 1995;**245**:171-183

[5] ASTM D 4406–93 (Reapproved 1998). Standard test method for creep of cylindrical rock core specimens in triaxial compression; 1998. pp. 1-5

[6] Aydan IT, Zbay U, Kwasniewski M, Shariar K, Okuno T, Zgenoglu A, et al. ISRM suggested methods for determining the creep characteristics of rock. Rock Mechanics and Rock Engineering. 2014;**47**:275-290

[7] Bieniawski ZT. Mechanism of brittle fracture of rock. International Journal of Rock Mechanics and Mining Sciences. 1967;**4**:395-430

[8] Dusseault MB, Fordham CJ. Time dependent behaviour of rocks. Comprehensive Rock Engineering. 1993; **3**:119-149

[9] Hemeda S. Engineering failure analysis and design of support system for ancient Egyptian monuments in Valley of the Kings, Luxor, Egypt. Geoenvironmental Disasters. 2018;**5**:12. DOI: 10.1186/s40677-018-0100-x

[10] Haupt M, Natau O. Uniaxial relaxation test on rock salt. In: Proceedings of the 2nd Conference on Mechanical Behaviour of Salt; Hanover; 1984. pp. 180-185

[11] Ladanyi B. Time dependent response of rock around tunnels. Comprehensive Rock Engineering. 1993; **2**:77-112

[12] Lama RD, Vutukuri VS. Handbook on Mechanical Properties of Rocks-Testing Techniques and Results. Vol. 21978. pp. 209-323

[13] Maranini E, Brignoil M. Technical note: Creep behaviour of a weak rock: Experimental characterization. International Journal of Rock Mechanics and Mining Sciences. 1999;**36**:127-138

[14] Nomikos. Supported axisymmetric tunnel within linear viscoelastic burger rocks. Rock Mechanics and Rock Engineering. 2011;**44**:553-564

[15] Pan YW, Dong JJ. Time-dependent tunnel convergence I. Formulation of the model. International Journal of Rock Mechanics and Mining Science and Geomechanics Abstracts. 1991;**28**:477-488

[16] Robertson EC. Creep of Solenhofen limestone under moderate hydrostatic pressure. Geological Society of America Memoirs. 1963;**79**:227-244

[17] Wang Y, Qi J, Yang C, Wei J. A study of nonlinear creep law in deep rocks. Rock and Soil Mechanics. 2005;**26** (1):117-121

[18] Zhu H, Ye B. Experimental study on properties of rock creep in saturation. Chinese Journal of Rock Mechanics and Engineering. 2002;**21**(12):1791-1796

[19] Zhang Z, Luo J. Study on creep properties of rock under step load. Chinese Journal of Rock Mechanics and Engineering. 2004;**23**(2):218-222

[20] Liu J, Yang C, Li X, Jiang D. Testing study on creep of coal rocks in the tunnel of Wankai speedway. Chinese Journal of Rock Mechanics and Engineering. 2004;**23**(22):3794-3798

[21] Parkin AK. Creep of rockfill (Part A). In: Maranhadas Neves E, editor. Advances in Rockfill Structure. London: Kluwer Academic Publishers; 1992. pp. 221-239

[22] Shen Z, Zhao K. Back analysis of creep deformation of rockfill dams. Journal of Hydraulic Engineering. 1998; **6**:1-6

[23] Guo X, Wang D, Cai X, Dong L. Rheological analysis of concrete faced rock-fill dam. Journal of Hydraulic Engineering. 1999;**11**:42-46

[24] Wang M, He X, Cheng Z. Current situation and prospect of studies on rheology property of course stuff. Rock and Soil Mechanics. 2003;**Suppl**: 451-454

[25] Liu Y, Li Y, Sun M. A new method of permeability test for loose rock. Ground Pressure and Strata Control. 2002;**19**(4):108-110

[26] Hemeda S. An integrated approach for the pathology assessment and protection of underground monuments in seismic regions. Application on some Greek-Roman Monuments in Alexandria, Egypt [PhD thesis]. Greece: Aristotle University of Thessaloniki; 2008

[27] Hemeda S, Pitlakis K. Serapeum temple and the ancient annex daughter library in Alexandria, Egypt: Geotechnical-geophysical investigations and stability analysis under static and seismic conditions. Engineering Geology. 2010;**113**:33-43

[28] Hemeda S, Pitilakis K, Bandis S, Papayianni I, Gamal M. The Underground monuments (Catacombs) in Alexandria, Egypt. In: Proceedings of the 4th International Conference on Earthquake Geotechnical Engineering; Thessaloniki, Greece; 25–27 June, 2007. pp. 715-738

[29] Bieniawski ZT. The geomechanics classification in rock engineering applications, In: Proceedings of the Fourth International Conference on Rock Mechanics, Vol. 2. Montreox: ISRM; 1979. pp. 41-48

[30] Aydan O, Akagi T, Ito T, Kawamoto T. Prediction of behaviour of tunnels in squeezing ground. Journal of JSCE. 1992;**440**(3–9):73-82

[31] Aydan O, Akagi T, Kawamoto T. The squeezing potential of rocks around tunnels: Theory and prediction. Rock Mechanics and Rock Engineering. 1993; **26**(2):137-163

[32] Aydan O, Kawamoto T. The stability assessment of a large underground opening at great depth. In: 17th International Mining Congress and Exhibition of Turkey (IMCET 2001); Ankara, Vol. 1; 2001. pp. 277-288

[33] Aydan O, Ulusay R. Geotechnical and geo-environmental characteristics of man-made underground structures in Cappadocia, Turkey. Engineering Geology. 2003;**69**:245-272

[34] Aydan O, Ulusay R, Kawamoto T. Assessment of rock mass strength for underground excavations. In: The 36th US Rock Mechanics Symposium; 1997. pp. 777-786

[35] Deere DU, Varde OA. General report, engineering geological problems related to foundations and excavations in weak rocks. In: Proceedings of the 5th International Association of Engineering Geology Congress, 1986. Vol. 4. Balkema; 1990. pp. 2503-2518

[36] Hemeda S. 3D finite element coupled analysis model for geotechnical

and complex structural problems of historic masonry structures: conservation of Abu Serga church, Cairo, Egypt. Heritage Science. 2019;7:6. DOI: 10.1186/s40494-019-0248-z

[37] Hemeda S. Geotechnical and geophysical investigation techniques in Ben Ezra Synagogue in Old Cairo area, Egypt. Heritage Science. 2019;7:23. DOI: 10.1186/s40494-019-0265-y

[38] Hemeda S, Pitilakis K. Geophysical Investigations at Cairo's Oldest, the Church of Abu Serga (St. Sergius), Cairo, Egypt. Research in Nondestructive Evaluation. 2017;**28**(3): 123-149. DOI: 10.1080/ 09349.847.2016.11439.91

[39] Bukovansky M, Richard DP, Week KR. Influence of slope deformations on the tombs in the valley of the kings, Egypt. In: Engineering Geology and the Environment-Proceedings-Symposium; Athens, Vol. 3; 1997. pp. 3077-3080

[40] Berest P, Charpentier JP, Blum PA, Gharbi H. Very slow creep tests on rock samples. International Journal of Rock Mechanics and Mining Sciences. 2005; **42**:569-576

[41] Brace WF, Kohlstedt DL. Limits of lithospheric stress imposed by laboratory experiments. Journal of Geophysical Research. 1980;**85**:6348-6352

[42] Brace WF, Paulding BW, Scholz CH. Dilatancy in the fracture of crystalline rocks. Journal of Geophysical Research. 1958;**71**:3939-3953

[43] Charles RJ. Fatigue of glass, I. Journal of Applied Physics. 1966;**29**: 1549-1560

[44] Adachi T, Oka F. An elasto-plastic constitutive model for soft rock with strain softening. International Journal for Numerical and Analytical Methods in Geomechanics. 1995;**19**:233-247

[45] Arces M, Nocilla, Aversa S, Lo Cicero G. Geological and geotechnical features of the "Calcarenite di Marsala". In: Evangelista, Picarelli, editors. Proceedings of the Geotechnics of Hard Soils-Soft Rocks. Rotterdam: Balkema; 1998

[46] Atkinson BK. Introduction to fracture mechanics and its geophysical applications. In: Atkinson BK, editor. Fracture Mechanics of Rocks. San Diego, CA: Academic; 1987. pp. 1-26

[47] Atkinson BK, Meredith PG. The theory of sub-critical crack growth with applications to minerals and rocks. In: Atkinson BK, editor. Fracture Mechanics of Rocks. San Diego, CA: Academic; 1987. pp. 111-166

[48] Aversa S. Mechanical behaviour of soft rock: Some remarks. In: Proceedings of the Workshop on "Experimental Characterization and Modelling of Soil and Soft Rocks. Napoli; 1991. pp. 191-223

[49] Aydan O, Tokashiki N, Kawamoto T. Microstructure models for porous rocks to jointed rock masses. In: Third Asia-Pacific Conference on Computational Mechanics; Seoul, Vol. 3; 1996. pp. 2235-2242

[50] Cecconi M, Viggiani G. Physical and structural properties of a pyroclastic soft rock. In: Evangelista, Picarelli, editors. Proceedings of the Geotechnics of Hard Soils-Soft Rocks. Rotterdam: Balkema; 1998

[51] Cooper MR. The mechanics of uncemented carbonate sands. Geotechnique. 1990;**40**(4):607-627

[52] Cripps JC, Coulthard JM, Forster A, Hencher SR, Moon CF. The engineering geology of weak rock. In: Proceedings of the 26th Annual Conference of the Engineering Group of the Geological Society, Leeeds (AA Balkema, Engineering Geology Special Publication, 8); 1993. 510pp

Chapter 3

Geotechnical Response Models for Steel Compliant Riser in Deepwater Clays

Hany Elosta

Abstract

The touchdown zone (TDZ) often proves to be a spot where cyclic bending stresses are the largest and is therefore a critical location for fatigue. Catenary steel compliant pipelines or risers (SCRs) are subject of much ongoing research, particularly with respect to their fatigue life, which is strongly influenced by seabed soil conditions in the TDZ. This chapter reviews the recent publications that might have an impact on the SCR-seabed interaction. The review starts by looking at the SCR general arrangement. Thereafter, the focus moves to the review of the recent research that studied the interactions between deepwater SCRs and the seabed. In addition, the review went over the analysis techniques of the SCR, including the modelling philosophy and models for geotechnical response. The research gap and the need for future research are identified.

Keywords: steel catenary pipeline, touchdown zone, rigid seabed, non-linear seabed model, lateral soil resistance, geotechnical response

1. Introduction

Catenary steel compliant risers (SCRs) have joined the riser family, building on the catenary equation that has assisted in creating bridges across the world. SCRs are commonly used with TLPs, FPSOs, semisubmersibles and spars, as well as fixed structures, compliant towers and gravity structures. SCRs have been accompanied by floating platforms since 1994 and were first used as export risers for Auger TLP in an 872 m water depth [1, 2]. Since then, SCRs have been employed with many applications. The number of SCRs is increasing quickly because of its simplicity, economic effectiveness, and well-known material properties. A free-hanging simple catenary riser is connected to a floating production vessel and the riser hangs at a prescribed top angle. The riser is free-hanging and gently curves down to the seabed at the touchdown point (TDP). At the TDP, the SCR pipe embeds itself in a trench and then evenly rises to the surface where it rests, and is effectively a static pipeline. SCRs may be described as consisting of three portions [3], as shown in **Figure 1**:

- Catenary zone, where the riser suspends in a catenary section
- Buried zone, where the riser pipe penetrates into a trench
- Surface zone, where the riser pipeline rests on the seabed.

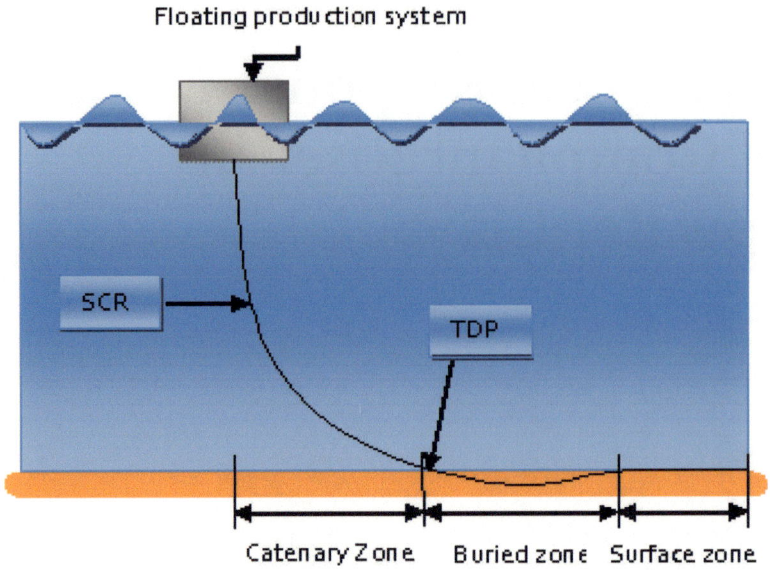

Figure 1.
General SCR arrangement.

A complex interaction between the SCR and seabed is experienced when the SCR is subjected to oscillatory motions. For SCRs, the most critical fatigue hotspot occurs in the TDZ. The SCR-seabed interaction is an essential key factor that should be considered in strength and fatigue assessment. How to precisely model this interaction response is still an issue and has been a hot field for academic research. A number of researches have been focused on understanding the soil-riser interaction. Better predictions of the SCR's fatigue life require an accurate characterisation model of seabed stiffness as well as a realistic description of the load/deflection curve. Therefore, this chapter gives a state-of-the-art review of the recent research on soil-riser interaction models. Briefly, a series of previous work associated with seabed-riser interaction mechanism and simulation models, as well as load/ deflection models, will be described and discussed.

2. SCR configuration design

The catenary riser length is estimated using simple geometric considerations, as following [4]:

$$L = \left(\frac{D - (MBR)A}{\cos\theta}\right) + (0.5\pi(MBR)A) \tag{1}$$

where L is the total length of the riser, D is the water depth, A is a factor depending on severity of environment (1.0 for mild environments and 1.2 for severe environments), θ is the riser top angle to vertical, typically between 10 and 25 degrees depending on severity of environment and water depth, and MBR is minimum bend radius based on 80% material yield strength. An additional riser pipeline length of approximately 750 m should be included to allow for TDP movement between near and far offset conditions as shown in **Figure 2**.

The SCR's static configuration must be determined before carrying out the dynamic analysis. The initial stage of any analysis of an SCR is the computation of

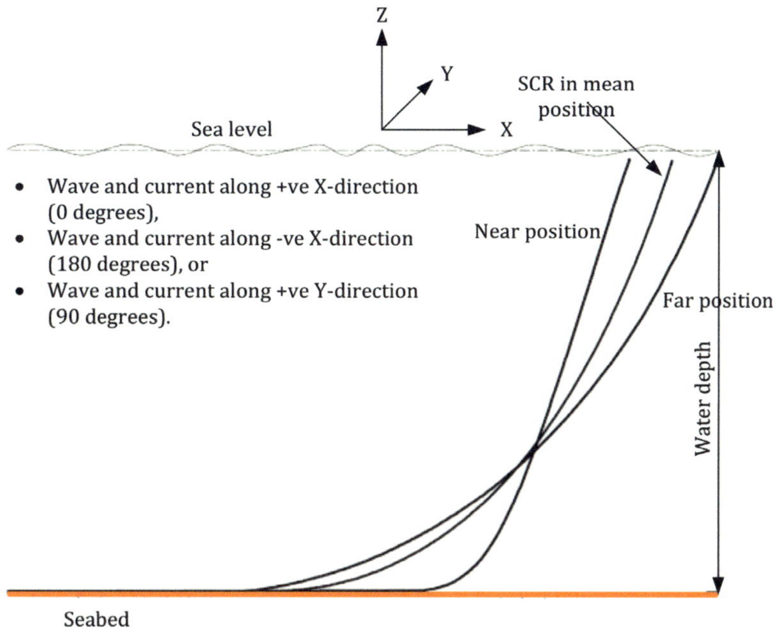

Figure 2.
Schematic of SCR configuration and vessel offsets.

its configuration under a set of static forces. The catenary equation gives a good first approximation for this, but in their basic formulation it only involves loads due to riser weight and assumes a riser pipe of zero bending stiffness, as presented before. However, the SCR static analysis is a large deflection non-linear behaviour problem with the influences of bending and tensional stiffness included. Therefore, many approaches have been developed to handle this problem using a combination of catenary equations and numerical techniques through iterative analysis. A review of existing approaches can be found in [5]. **Figure 3** shows an example of static configuration of an SCR in a 910 m water depth with a hang-off angle of 20° and a 273 mm outer diameter connected to the floating platform (FP) in the zero mean offset position (i.e., the FP is in its initial position without drifting in any direction), which is calculated using OrcaFlex/finite element analysis (FEA) software.

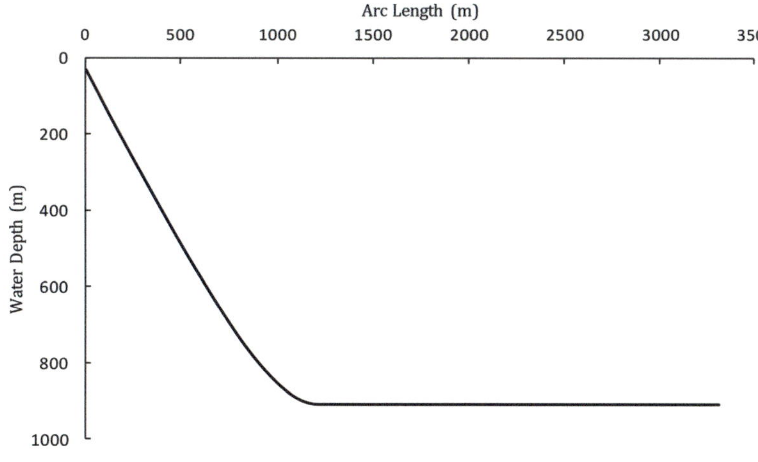

Figure 3.
Static configuration of SCR model.

Generally, SCRs have a limited amount of additional pipeline length available to accommodate the FP motions. Alterations in the catenary suspended length are obtained by the riser either being picked up or laid down on the seabed. Limitations are approached when either the SCR tension at the FP becomes too great as the FP drifts away from the TDP (far load case, as shown before in **Figure 2**) or when the bending stresses near the seabed become too great as the FP drifts towards the touchdown point (near load case). SCRs are less appropriate for FPSO applications where vessel offsets are considerably higher. **Figure 4** shows the effect of the horizontal vessel offset on the horizontal projection of the TDP. While the top of the SCR has the highest tension and lowest bending moment, the TDP has the lowest tension and the highest bending moment. The maximum bending stress and effective tension along the SCRs' arc length and the horizontal projection of the TDP due to the vessel offsets are presented in **Figures 5** and **6**, respectively.

The vessel offset governs the maximum bending stress at the TDP and also the maximum tension at the riser's top end. In the left region of **Figure 6**, where the

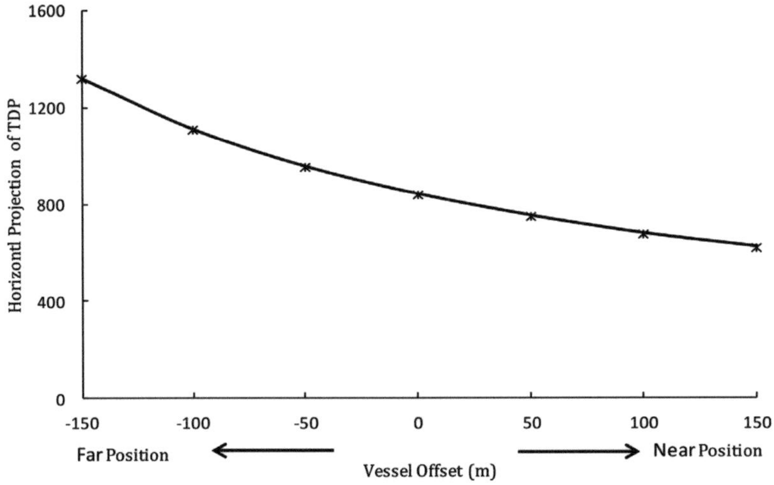

Figure 4.
Effect of the vessel offsets on the horizontal projection of the TDP.

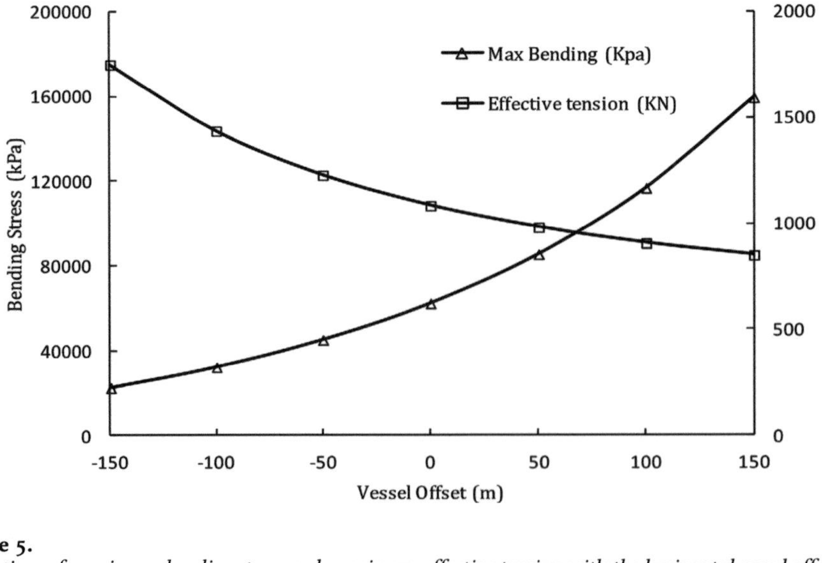

Figure 5.
Alterations of maximum bending stress and maximum effective tension with the horizontal vessel offset.

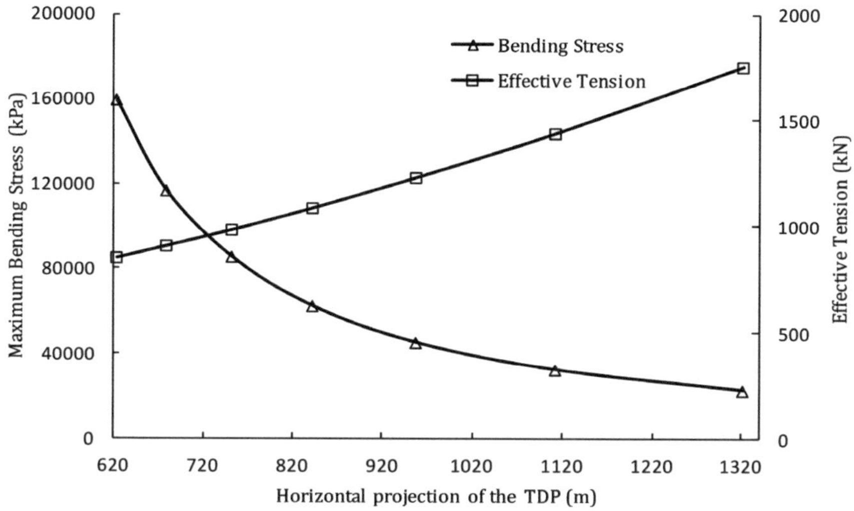

Figure 6.
Alterations of maximum bending stress and effective tension with the horizontal projection of the TDP.

vessel drifts towards the TDP (near load case), the bending stress at the TDP is increased rapidly within small change in the vessel offset. In the intermediate region, the bending stress and tension are slightly increased with the vessel offset. In the right region, where the vessel drifts away from the TDP (far load case), the bending stress slightly decreased, while the tension increased. Therefore, the conclusion from these results is that the vessel mean position should be offsetting the TDP with a roughly distance of 0.75–1.5 of the water depth. Furthermore, the catenary equation is a simple implementation tool to figure out the load distribution, geometric properties and static loads on an SCR. The specialist non-linear/FEA is implemented for the SCR design to tackle the complex nature of non-linear, large deflection behaviour of SCRs and to be post-processed quickly. The evaluation of the forces and behaviour of SCRs in the TDZ need more sophisticated methods.

3. Models for seabed response

3.1 Problem description of an SCR pipe embedment

SCR pipe penetration is defined as the depth of penetration of the pipe invert (bottom of pipe), relative to the undisturbed seabed as shown in **Figure** 7. Pipe penetration affects the riser pipe-seabed contact area, which subsequently affects the axial and passive soil resistance against the riser. Consequently, the passive soil resistance influences the lateral breakout force. Heave of seabed soil during embedment increases the local penetration of the SCR pipe by raising the soil surface level against the shoulders of the pipe. The typical geometry of heave produced during vertical embedment of an SCR pipe is such that the nominal penetration is approximately 50% less than the local embedment relative to undisturbed seabed surface [6].

The SCR-seabed interaction response characteristic is a highly non-linear phenomenon. It is important not to restrict the modelling of this interaction to a linear seabed model approximation and the riser analysis techniques must be improved by refining the riser-seabed interaction [7]. SCR-seabed interaction modelling should involve vertical and lateral soil responses to the cyclic loading oscillations of the SCR in the TDZ, which can cause trenching and dynamic embedment of the SCR into the

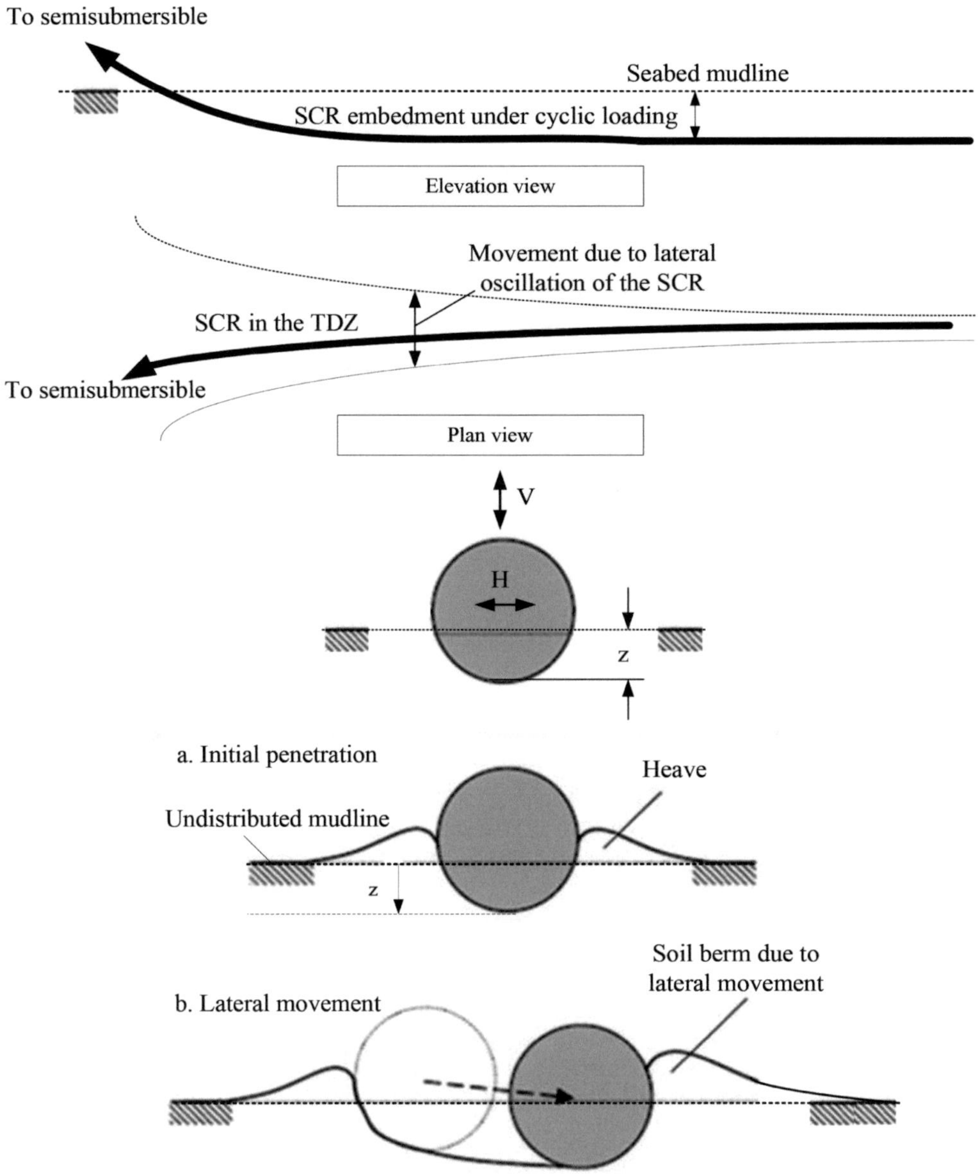

Figure 7.
Schematic illustration of the SCR-seabed interaction in the TDZ.

seabed. A typical schematic illustration of the SCR-seabed interaction and trench formation in the TDZ are shown in **Figure** 7.

3.2 SCR/seabed vertical interaction

The application of SCR systems has increased with the progressive development of hydrocarbon production further offshore and into deeper waters. The SCR-soil interaction at touchdown with the seabed is a major key factor for SCRs. An SCR is subjected to oscillatory motions from the host vessel and wave action. Therefore, the SCR experiences a complex interaction between the riser and seabed in the touchdown area, and deep trenches thus cut into the seabed in the buried zone beyond the TDP [8–10].

An appropriate SCR-seabed interaction model must be used. The TDZ is one of the key locations where the fatigue damage happens. The sophistication of the interaction model depends on the type of analysis and accuracy required. These interaction models vary from a simple rigid seabed with soil friction coefficients to more sophisticated ones, including vertical and lateral stiffness, friction and suction.

3.2.1 Rigid and elastic seabed

A potential fatigue damage of the SCR in the TDZ is related to maximum bending stress in the SCR, which relies on the soil stiffness of the seabed and the motions of the SCR. For example, the SCR on a soft seabed will have reduced bending stresses when a load is applied, while the one on a rigid seabed will have more critical bending stresses. A rigid surface generally contributes a conservative result, since it is unyielding, while the non-linear soil model is a better approximation of a seabed. Extreme offsets of the floating production unit with soft seabed model may then give higher stresses than those calculated on rigid seabed stiffness, since the catenary pipeline must be broken out of the seabed soil and high suction forces must be overcome. **Figure 8** shows a schematic of an SCR close to the TDP with the forces acting on a rigid seabed. The shear force F in the near horizontal segment close to the TDP is given by:

$$F = \frac{dM}{dx} = \frac{d}{dx}\left[EI\left(\frac{d^2y}{dx^2}\right)\right] = EI\left(\frac{d^3y}{dx^3}\right) \tag{2}$$

then $d^3y/dx^3 = (w/H)ke^{-kx}$ and the shear force close to the TDP is thus given by:

$$F = EI\left(\frac{w}{H}\right)ke^{-kx} \tag{3}$$

The bending moment at the TDP diminishes from the catenary bending moment to zero, and there is a concentrated reaction force. Since $k = \sqrt{H/EI}$, the TDP shear force that is transmitted to the soil [2, 11] is given by:

$$F_{TDP} = w\sqrt{EI/H} \tag{4}$$

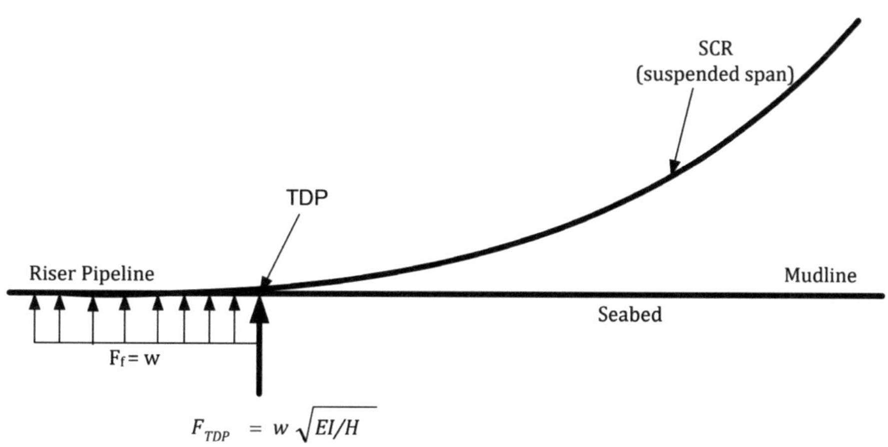

Figure 8.
Configuration of SCR close to TDP with a rigid seabed.

where F_{TDP} is the concentrated reaction at the TDP, assuming a rigid surface seabed, F_f is the reaction to the pipe resting on the seabed, as shown in **Figure 8**.

For the elastic soil response, configuration of an SCR close to the TDP is shown in **Figure 9** by representing a seabed with a linear elastic model. The curvature in the surface zone (i.e., the pipeline is resting on the seabed) is zero. In the TDZ, the riser's pipe is resting on linear elastic foundations. The solution for a beam element resting on an elastic foundation can be found in [12, 13], who introduced solutions that can be implemented for SCR-seabed interaction.

3.2.2 Non-linear load/deflection model

The current practice for the FEA of SCR-seabed interaction response is to model this interaction as structural soil springs [10] by using the developed models for buried pipelines and strip foundation theory. The conventional modelling of riser-seabed interaction use the non-linear elastic load/deflection curves, as described in [14]. Since the resistance force does not exceed the friction resistance limit (μV), the soil spring has a constant stiffness coefficient, *K*. The load/deflection model has zero resistance force at zero displacement, as the pipe displacement is increasing the resistance force also increases linearly until the peak seabed resistance is approached. When the seabed friction exceeds the limit friction force, the resistance force becomes constant with changing pipe displacement (large displacements occur without a further increase in the friction resistance force) and the spring stiffness becomes zero (i.e., slip occurs). The maximum seabed resistance load is given by the backbone curve [15], which corresponds to virgin penetration of the riser pipe into the seabed.

Linear soil stiffness can be used by FEA codes to model the non-linear riser-seabed interaction curves. Linear soil stiffness is defined as the ultimate bearing load divided by the riser pipe displacement, as given below:

$$K = \frac{V}{\Delta} \tag{5}$$

where *K* is the soil stiffness per unit length; *V* is the force per unit length; Δ is the riser pipe displacement. Different approaches are used to characterise the linear seabed stiffness, such as secant, tangent and Young's modulus stiffness, for more details see Barltrop et al. [16]. Herein, the secant stiffness type is considered because it is more stable than the tangent stiffness approach and is being used to model the load/deflection curve. The secant stiffness is the average stiffness between two points, typically the origin and the point in question, see **Figure 10**. Static seabed

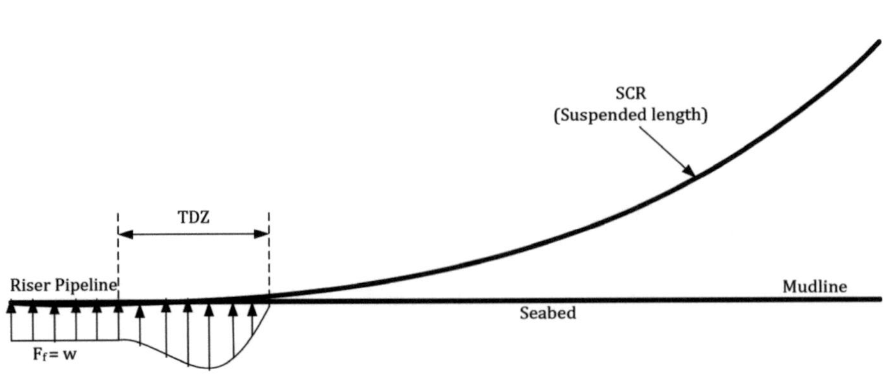

Figure 9.
SCR's configuration close to TDZ with linear elastic seabed.

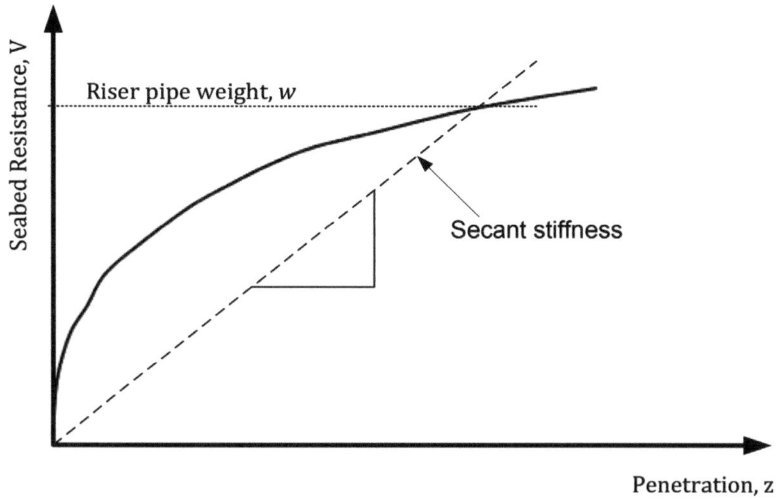

Figure 10.
Static and secant stiffness for non-linear seabed V-z *model.*

stiffness is using a linear stiffness to represent the backbone curve in riser-seabed interaction analysis. It is assumed that the riser will penetrate into the seabed until the bearing load equals the submerged riser pipe weight; $w = V_u$ where w is the submerged riser pipe weight, and V_u is the ultimate bearing load.

The non-linear soil model is recently developed and based on a hyperbolic secant stiffness formulation proposed by Bridge et al. [17], Aubeny et al. [18], and Randolph and Quiggin [19]. The non-linear seabed model is more sophisticated than the linear model, as it models the non-linear hysteretic behaviour of the seabed in the vertical direction, including modelling of suction effects when the SCR rises up sufficiently. The model uses data such as the pipe diameter, the seabed soil shear strength profile with depth and the soil density as its primary sources. Different functions are used for the initial penetration, for uplift and for re-penetration, whilst the function parameters are updated each time a penetration reversal occurs. This enables the model to capture the hysteretic behaviour of the seabed soil response and the increasing penetration of the pipe under cyclic loading in the vertical plane.

The typical *V-z* curve patterns [17, 19], as shown in **Figure 11**, of pipe-soil interaction are produced by laboratory model experiments [15] of vertically loaded horizontal pipes in weak sediment. These curves can be divided into four different paths. The pipe-soil interaction process is described and the depiction of the development of the interaction curve is given in **Figure 11**, associated with the uplift/re-penetration cycle. If the riser pipe continues to experience oscillatory loading movement, the *V-z* interaction curve will repeat the loop enclosed by path 1-2-3-1 under the assumption of a non-degradation model.

3.3 SCR/seabed lateral interaction

One of the main issues encountered with the use of the SCR is the large lateral movements on the seabed due to the FP motions and marine environment. Thus, better understanding of the lateral soil resistance to SCR pipe movements must be considered for SCR design. Many researchers had focused on studying and investigating the pipe-seabed lateral interaction response [6, 20–24]. Three different approaches [25, 26] can be considered for determining the lateral soil resistance of partially embedded pipelines:

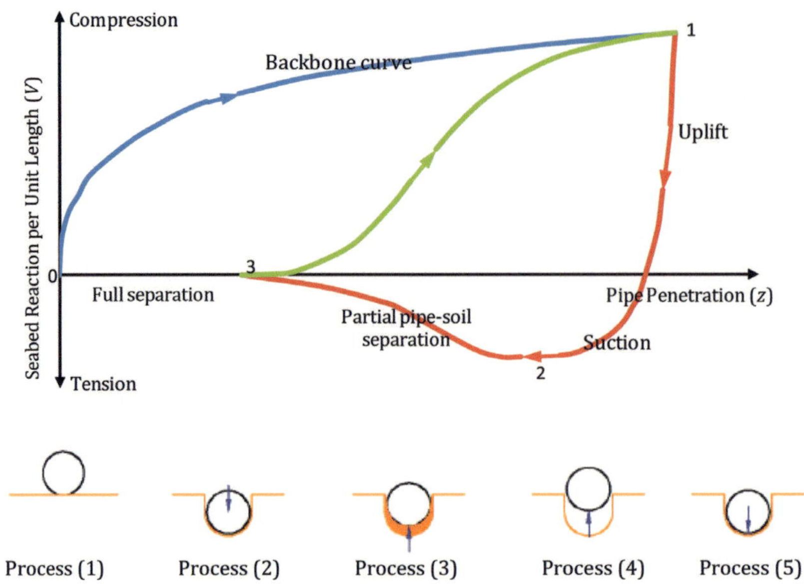

Figure 11.
Depiction of typical V-z *behavior [7].*

- A single friction factor "Coulomb friction model" approach, where the lateral soil resistance is related to the submerged weight of the pipeline and the soil type. This approach is fairly simplified, as it does not pertain to pipe embedment;
- A two-component model, where the lateral soil resistance consists of a sliding resistance component and a lateral passive pressure component [20, 23, 27]; and
- A plasticity model approach: Zhang et al. initially developed the plasticity model for calcareous sand and clays [28]. However, the clay's model is established on the behaviour of shallow flat footings in which a large lateral movement does not occur.

Therefore, the Coulomb friction approach and the two-component soil resistance models for the assessment of SCR global response are presented in this chapter.

3.3.1 Coulomb friction "Bilinear" model

Present industry procedure is to evaluate the soil resistance with a Coulomb friction model, as shown in **Figure 12**, which expresses the lateral resistance as the product of the effective submerged pipeline vertical force (submerged pipe weight minus hydrodynamic lift force) and a soil friction coefficient that depends solely on soil type. Recommended values of the Coulomb friction coefficient, μ, lie in the range 0.2–0.8, while the displacement to mobilise this resistance is typically 0.1 pipe diameters [20, 23, 24, 29].

3.3.2 Tri-linear pipe-seabed interaction model

The experiment results show that the pipe-soil lateral motion is far more complicated than simple coulomb friction. An improved model is essential in order to mimic

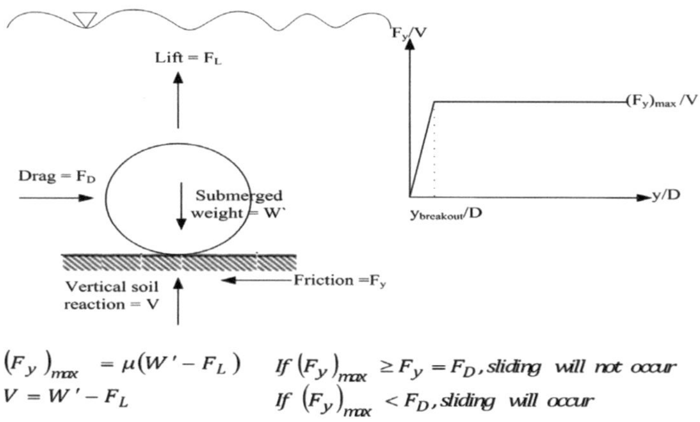

Figure 12.
Coulomb friction model analysis.

the effects of soil strength and the load history of the catenary pipeline as well as the associated pipe embedment on the lateral seabed soil resistance. The improved empirical model utilises two components to predict the seabed resistance to lateral pipeline movements, resulting in the improved so-called "two-component model."

The two-component model uses an empirical formula to assess the soil resistance to lateral pipeline motions. The first component depends on the vertical pipe weight (pipe weight minus hydrodynamic lift force) and imitates the sliding resistance of the pipeline along the soil surface, while the second component depends on the pipe penetration and soil strength.

Generally, the two-component models are based on empirically fitting laboratory test data. A summary of some of the proposed formulas is given in **Table 1**. The peak lateral soil resistance is a key parameter for the on-bottom pipeline movement. Several reported methods [20, 23, 27, 30] have been published for the assessment of the lateral soil resistance. These determined resistances were then compared with the results of the available pipe model tests.

Figure 13 shows the lateral load response from step 0 to 3, characterised as follows [31]:

(0–1)	First load breakout, with elastic response characterised by the mobilisation displacement and a peak that is dependent on the initial pipe embedment;
(1–2)	Suction release phase and elevation correction, depending on initial pipe embedment;
(2–3)	Steady state of residual friction.

3.4 SCR/seabed axial interaction

The axial soil resistance for SCR movement is typically modelled using the Coulomb friction model, which is adopted to evaluate the axial resistance of a partially embedded riser pipe, and is expressed as [25, 32]:

$$F_x = \mu_A W_s \tag{6}$$

where F_x is the axial soil resistance and μ_A is the coefficient of axial coil friction. The typical values for axial friction have been reported to vary between 0.2 and 0.5 for clay soil [33].

Author	Lateral soil resistance formulas	Comments
Wagner et al. [23]	$F_y = \mu(W' - F_L) + \beta S_u A/D$ F_y = lateral soil resistance μ = sliding resistance coefficient W' = submerged pipe weight F_L = hydrodynamic lift β = empirical soil passive resistance coefficient S_u = undrained shear strength of the clay A = 0.5 × embedded area	Monotonic $\mu = 0.2$ $\beta = 39.3$ Cyclic (oscillations below the monotonic breakout load [<static failure]) Penetration × 2 $\beta = 31.4$ Cyclic (large displacement oscillations) Penetration × 2.5 $\beta = 15.7$
Brennodden et al. [20]	$F_y = \mu(W' - F_L) + F_R$	$\mu = 0.2$ F_R calculated considering accumulated energy
Verley and Lund [27]	$F_y = F_C + F_R$ $F_y = \mu(W' - F_L) + F_R$ $F_R = 4.13DS_u\left(\frac{S_u}{\gamma D}\right)^{-0.392}\left(\frac{z}{D}\right)^{1.31}$	Clays (S_u < 70 kPa) $\mu = 0.2$
Bruton et al. [30]	$(F_y)_{dimensionless} = \mu v + \frac{3}{\sqrt{\frac{S_u}{\gamma' D}}}\frac{z_{init}}{D}$ $(F_y)_{dimensionless} = \frac{F_y}{S_u D}$ $\frac{z_{init}}{D}$ = normalised initial pipe penetration $v = \frac{V}{DS_u}$ normalised vertical load due to the effective pipe weight $\frac{(f_y)_{res}}{v} = 1 - 0.65\left[1 - \exp\left(-\frac{1}{2}\frac{S_u}{\gamma' D}\right)\right]$	Soft clays (0.15 < S_u < 8 kPa) $\mu = 0.2$

Table 1.
Lateral resistance models of partially embedded pipelines in soft clay.

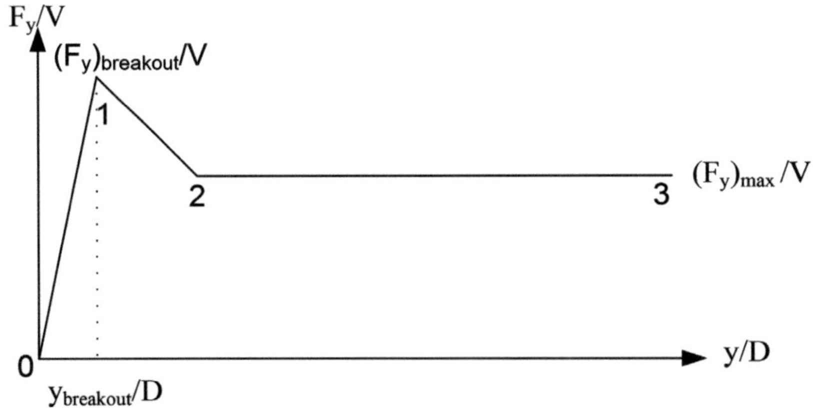

Figure 13.
Schematic of the tri-linear soil resistance model.

4. Summary of SCR-seabed interaction models technique

A number of scientific papers have been published on the study of soil-riser interaction. **Table 2** details the considerable diversity in the level of sophistication used in the analysis of SCR-seabed interaction response in a representative set of studies published in the last 25 years, with five areas previously highlighted as commonly conservative and broken into components of increasing degree of sophistication and accuracy. **Table 2** have been addressed for appropriate modelling of the physical SCR-seabed interaction process and graded with a three levels ranging from α which represents state-of-the-art practice, β a compromise method and γ a conservative method. General and brief discussion on the components presented in **Table 2** are as follows:

SCR-seabed interaction models: The majority of experimental studies carried out in recent years have presented the non-linear behaviour of SCR-seabed interaction and trenching effects in the TDZ. An SCR in the TDZ was recently identified as a fatigue hotspot that substantially increased fatigue damage under the SCR-seabed interaction phenomenon. Better understanding of the significance of SCR-seabed interaction behaviour and soil properties improves the fatigue life prediction in the TDZ. Most of the existing riser models unrealistically simplified SCR-seabed interaction behaviour by assuming a rigid or linearly elastic seabed. Trench formation and trench deepening have also significant influence on SCR-seabed response.

Pipe-seabed interaction models: Experimental model tests and analytical models of vertically loaded horizontal pipes in clay sediment provided valuable information for better awareness of SCR-seabed interaction in the TDZ. These experimental and analytical data produce the general load/deflection curve for pipe-seabed interaction and necessary information for validation of *V-z* model and determination of geotechnical parameters used in the model.

Cyclic loading: SCR-seabed interaction processes should cover vertical and lateral responses to cyclic loading. Movement and oscillation of the SCR in the TDZ will cause trenching and dynamic embedment of the SCR into the seabed. Seabed stiffness degradation due to cyclic oscillations has a significant influence on the behaviour of an SCR in the TDZ, and especially on the SCR's strength and fatigue performance. After the seabed soil approaches the maximum strength throughout the applied cyclic loading, the seabed soil tends to lose strength and stiffness with the increase in plastic embedment during cyclic oscillations. The seabed soil stiffness degradation mechanism comprises stiffness reduction presented in uplift, suction, and separation as well as the re-penetration process. The degradation of soil stiffness with cyclic loading is best captured by the non-linear seabed model.

Wave loading: The use of regular wave theories does not adequately represent wave loading on SCRs. However, when used, the level of sophistication in random wave loading is highly variable. For example, the length of simulation used to estimate response levels differs widely. Most studies use simplifying assumptions due to the extensive computational time needed to perform random time domain simulation properly.

Analysis (coupled vs. uncoupled): SCRs have a relative effect on the motions of a floating unit, which in turn affects the SCR fatigue life. In a coupled analysis, the floating unit and SCR are modelled together including their mass, stiffness and damping. Coupled analysis is computationally demanding, especially for robust finite element mesh size. In uncoupled analysis, platform wave frequency is computed in separate models by different programs. Once the floating unit motions are obtained, either from coupled or uncoupled analysis, they are imported as input into the riser analysis software for the uncoupled riser analysis. In uncoupled analysis, the riser is considered to have no effect on the platform at its top. These effects are usually negligible, and an uncoupled analysis is adequate.

5. Further research

The literature review introduced in this chapter reveals that the SCR motion at the TDP is predominantly lateral, vertical and cyclic in nature. SCRs are the subject

Author	Year	SCR-seabed interaction models					Pipe-seabed interaction models					Cyclic loading		Wave loading		Analysis	
		Model tests	Analytical models				Model tests	Analytical models									
			γ	β	α	α		γ	β	α	α	α	α	β	α	β	α
			Rigid seabed	Linear seabed	Non-linear seabed	Trenching effects		Rigid seabed	Linear seabed	Non-linear seabed	Trenching effects	Vertical	Lateral	Regular	Irregular	Uncoupled	Coupled
Sharma and Aubeny [34]	2011				√	√						√			√	√	
Cao [35]	2010			√													√
Cardoso and Silveira [36]	2010						√						√			√	
Hodder et al. [37]	2010						√					√				√	
Jin et al. [38]	2010						√						√			√	
Kimiaei et al. [39]	2010				√									√		√	
Nakhaee [40, 41]	2008 and 2010									√	√	√		√		√	
Aubeny et al. [18, 42]	2006 and 2009									√	√	√				√	
Randolph and Quiggin [19]	2009									√	√	√			√	√	
Oliphant et al. [43]	2009						√						√			√	
Bruton et al. [6, 30]	2006 and 2008						√						√			√	
Clukey et al. [44]	2008						√					√				√	
Palmer [11]	2008		√													√	
Sen [45, 46]	2008			√													√
Xia et al. [47]	2008			√											√		√

Author	Year	SCR-seabed interaction models					Pipe-seabed interaction models					Cyclic loading		Wave loading		Analysis	
		Model tests	Analytical models				Model tests	Analytical models									
			γ	β	α	α		γ	β	α	α	α	α	β	α	β	α
			Rigid seabed	Linear seabed	Non-linear seabed	Trenching effects		Rigid seabed	Linear seabed	Non-linear seabed	Trenching effects	Vertical	Lateral	Regular	Irregular	Uncoupled	Coupled
Akpan et al. [48]	2007		√														√
Karunakaran et al. [9]	2005			√											√		√
Bridge et al. [17]	2004											√					
Giertsen et al. [49]	2004						√		√	√	√	√					
Bridge et al. [3]	2003	√					√					√					√
Langner [50]	2003		√												√		√
Bridge and Willis [51]	2002	√										√					√
Thethi and Moros [10]	2001	√										√					√
Willis and West [52]	2001	√										√					√
Pesce et al. [53]	1998			√													
Verley and Lund [27]	1995						√						√				
Hale et al. [54]	1992						√					√	√				
Dunlap et al. [15]	1990						√					√					
Brennodden et al. [20]	1989						√						√				
Morris et al. [22]	1988						√						√				
Wagner et al. [23]	1987						√						√				

Table 2.
Level of complexity used in SCR-seabed interaction technique.

of much ongoing research, particularly with respect to fatigue and interaction with the seabed at the TDP. The current SCR's analysis is performed using simplified pipe-seabed interaction models and disregards the SCR's embedment into the seabed as well as soil suction effects in the TDZ; this will affect the predicted SCR response. Previous experiments showed that the soil suction effect can increase the bending stress of SCRs in the TDZ. The predominant offshore soil condition in a deepwater environment is soft clay soil with small undrained shear strength. Field observations have introduced that the trench is a common feature due to the SCR pipe penetration into the seabed. However, there are few published literatures that investigate the trenching effects of the riser pipe in the TDZ on the SCR's dynamic structural behaviour and fatigue performance.

Seabed stiffness degradation due to cyclic motion is an important parameter in order to estimate the SCR fatigue in the TDZ. This aspect is not well investigated, and the seabed is traditionally not properly modelled in the current SCR fatigue analysis. Existing literature has introduced that fatigue damage is highly sensitive to the soil model utilised in the fatigue estimation calculation. The seabed non-linear model, to simulate the SCR-seabed interaction, has been shown to be more sophisticated compared to those SCR-seabed interactions with linear soil springs.

It can be concluded from the summary of models presented in the existing literature survey that:

- Although a linear seabed model is the common modelling for seabed response [9, 47, 55], which is too simplified to capture the nature of SCR-seabed interaction, SCR-seabed interaction is a considerably non-linear phenomenon. For better understanding of SCR behaviour and reliable prediction of the fatigue life in the TDZ, a numerical study and analysis of SCRs with vessel motions should be performed;

- Fatigue performance assessment of the SCR in the TDZ remains a serious design challenge for SCR behaviour. Despite some research papers presenting a reduction in fatigue damage [34, 40, 50] due to riser embedment in the TDZ, other studies have proposed an increase in fatigue damage [7, 49]. These confounding results due to different geotechnical parameters have been imposed with trenches. The trench deepening, gradual embedment of the riser and soil stiffness have an important influence on the SCR's fatigue life in the TDZ. Furthermore, the soil parameters used in riser-seabed analysis can have a significant effect on fatigue life of SCRs. Therefore, SCR's fatigue damage in the TDZ is a critical design aspect where geotechnical consideration becomes important; and

- Although lateral movements of the SCR can influence the riser performance, as suggested by [42, 49], the adduced SCR-seabed interaction analytical models regard only the vertical SCR motions and neglect lateral soil friction, as presented by [40, 41].

The aforementioned research gap points are subjected to ongoing research and investigations and being tackled by the authors.

6. Conclusion

The interested engineer or researcher will find here the necessary background on the geotechnical interaction model for SCR issue, and then will be able to proceed with the research literature. In this chapter, the main objective was to explore the various modelling approaches used in recent studies towards better clarification of the response behaviour of pipe-soil interaction under cyclic motions.

The seabed response due to riser loading and the trench formation phenomenon are of great significance for safe and economic riser design. Current studies of SCR technology focused on better understanding of the TDZ and its interaction with the seabed soil. The soil-riser interaction involves a number of complexities, including non-linear soil behaviour, trench width and depth variability and softening of the seabed soil under cyclic loading. The seabed-riser interaction modelling allows the effect of physical phenomena, such as lateral resistance, soil suction forces and vertical seabed stiffness on the SCR performance to be identified and investigated. Non-linear seabed-riser model interaction will determine the influence of the seabed response model on SCR fatigue. A small change in seabed stiffness can result in a small change in bending stress, but this causes a significant change in fatigue life. Therefore, the need for seabed-riser interaction modelling to be as realistic as possible is evident. A comprehensive review of the recent studies on the SCR-seabed interaction was introduced.

After reviewing the different parts of literature relevant to this study, some of the required knowledge to be used in the current and future research is acquired and some other existing gaps in the field are identified. This chapter has presented a review of the state-of-the-art SCRs with seabed interaction and analysis techniques. It has also discussed the existing theories for modelling SCR-seabed interaction with detailed explanation of currently used methods for evaluating the SCR structural performance in the TDZ. The research gap addressed in this chapter is under the investigation and ongoing research by the author.

Author details

Hany Elosta
Subsea SURF Lead, TechnipFMC, Lysaker, Norway

*Address all correspondence to: hanyelosta@gmail.com

References

[1] Mekha BB. New frontiers in the design of steel catenary risers for floating production systems. Journal of Offshore Mechanics and Arctic Engineering. 2001;**123**:153

[2] Sparks C. Fundamentals of Marine Riser Mechanics: Basic Principles & Simplified Analysis. PennWell; 2008

[3] Bridge C, Howells H, Toy N, Parke G, Woods R. Full scale model tests of a steel catenary riser. In: Proceding of International Conference on fluid Structure Interaction; Cadiz, Spain. 2003

[4] Hatton S. Update on the design of steel catenary riser systems. In: Society of Underwater Technology; Newcastle, UK. 1999

[5] Patel M, Seyed F. Review of flexible riser modelling and analysis techniques. Engineering Structures. 1995;**17**:293-304

[6] Bruton DAS, White DJ, Carr M, Cheuk JCY. Pipe-soil interaction during lateral buckling and pipeline walking—The safebuck JIP. In: Offshore Technology Conference; 5-8 May 2008; OTC 19589, Houston, Texas

[7] Leira BJ, Passano E, Karunakaran D, Farnes KA, Giertsen E. Analysis guidelines and application of a riser-soil interaction model including trench effects. In: 23rd International Conference on Offshore Mechanics and Arctic Engineering, OMAE2004-51527. Vancouver, British Columbia, Canada: ASME; 2004

[8] Karunakaran D, Lund KM, Nordsve NT. Steel catenary riser configurations for north sea field developments. In: Offshore Technology Conference; OTC 10979, Houston, Texas. 1999

[9] Karunakaran D, Meling T, Kristoffersen S, Lund K. Weight-optimized SCRs for deepwater harsh environments. In: Offshore Technology Conference; OTC 17224; 2-5 May 2005; Houston, TX, USA

[10] Thethi R, Moros T. Soil interaction effects on simple catenary riser response. In: Proceeding Conference on Deepwater Pipeline & Riser Technology; Houston, Texas. 2001

[11] Palmer A. Touchdown indentation of the seabed. Applied Ocean Research. 2008:235-238

[12] Hetényi M, Michigan U o. Beams on Elastic Foundation. The Univ. of Michigan Press; 1946

[13] Jones G. Analysis of Beams on Elastic Foundations: Using Finite Difference Theory. Inst of Civil Engineers Pub; 1997

[14] Chen WF, Han DJ. Tubular Members in Offshore Structures. Pitman Publishing Inc.; 1985

[15] Dunlap WA, Bhojanala RP, Morris DV. Burial of vertically loaded offshore pipelines in weak sediments. In: Offshore Technology Conference, OTC 6375; 7-10 March 1990; Houston, Texas

[16] Barltrop NDP, Adams AJ, Hallam M. Dynamics of Fixed Marine Structures. Oxford: Butterworth-Heinemann; 1991

[17] Bridge C, Laver K, Clukey E. Steel catenary riser touchdown point vertical interaction models. In: Offshore Technology Conference, OTC 16628; Houston, Texas. 2004

[18] Aubeny C, Biscontin G, Zhang J. Seafloor interaction with steel catenary risers. In: Texas: Offshore Technology Research Center, Texas A&M University College Station, Houston, Final Project Report to Minerals

Management Service, OTRC Library Number 9/06A173. 2006

[19] Randolph M, Quiggin P. Non-linear hysteretic seabed model for catenary pipeline contact. In: 28th International Conference on Ocean, Offshore and Arctic Engineering, OMAE2009-79259. Honolulu, Hawaii, USA: Proceedings of the ASME 2009; 2009

[20] Brennodden H, Lieng J, Sotberg T, Verley R. An energy-based pipe-soil interaction model. In: Offshore Technology Conference; Houston, Texas, USA. 1989

[21] Cheuk C, Bolton M. A technique for modelling the lateral stability of on-bottom pipelines in a small drum centrifuge. In: Physical Modelling in Geotechnics—6th ICPMG; London. 2006.

[22] Morris DV, Webb RE, Dunlap WA. Self-burial of laterally loaded offshore pipelines in weak sediments. In: Offshore Technology Conference; OTC 5855, Huston, Texas. 1988

[23] Wagner D, Murff J, Brennodden H, Sveggen O. Pipe-soil interaction model. In: Offshore Technology Conference; 27-30 April 1987; Houston, Texas, USA. OTC

[24] White DJ, Cheuk CY. Modelling the soil resistance on seabed pipelines during large cycles of lateral movement. Marine Structures. 2008;**21**:59-79

[25] Cathie D, Jaeck C, Wintgens J-F Ballard J-C. Pipeline geotechnics—State of the art. In: Proc. International Symposium on Frontiers in Offshore Geotechnics (ISFOG); Perth, Australia. 2005

[26] Dendani H, Jaeck C. Flowline and riser: Soil interaction in plastic clays. In: Offshore Technology Conference, OTC 19261; 5-8 May 2008; Houston, Texas, USA

[27] Verley R, Lund KM. A soil resistance model for pipelines placed on clay soils. In: American Society of Mechanical Engineers. 1995. pp. 225-225

[28] Zhang J, Stewart D, Randolph M. Modeling of shallowly embedded offshore pipelines in calcareous sand. Journal of Geotechnical and Geoenvironmental Engineering. 2002; **128**:363

[29] Lyons C. Soil resistance to lateral sliding of marine pipelines. In: Offshore Technology Conference; 1973; OTC 1876, Texas, USA

[30] Bruton D, White D, Cheuk C, Bolton M, Carr M. Pipe/soil interaction behavior during lateral buckling, including large-amplitude cyclic displacement tests by the safebuck jip. In: Offshore Technology Conference; Houston, USA. 2006

[31] Bruton D, Carr M, Crawford M, Poiate E. The safe design of hot on-bottom pipelines with lateral buckling using the design guideline developed by the safebuck joint industry project. In: Deep Offshore Technology Conference; Espirito Santo, Brazil. 2005

[32] Oliphant J, Maconochie A. The axial resistance of buried an unburied pipelines. In: Offshore Site Investigation and Geotechnics, Confronting New Challenges and Sharing Knowledge. 2007

[33] DNV 2007. Dnv-rp-f113: Recommended practice pipeline subsea repair. Det Norske Veritas

[34] Sharma P, Aubeny C. Advances in pipe-soil interaction methodology and application for SCR fatigue design. In: Offshore Technology Conference; Houston, USA. 2011

[35] Cao J. Fatigue analysis technique of deepwater steel catenary risers. In: 29th International Conference on Ocean, Offshore and Arctic Engineering,

OMAE 2010; 6-11 June 2010. Shanghai, China: ASME

[36] Cardoso C, Silveira R. Pipe-soil interaction behavior for pipelines under large displacements on clay soils—A model for lateral residual friction factor. In: Offshore Technology Conference; OTC 20767, Houston, Texas. 2010

[37] Hodder MS, Byrne BW. 3d experiments investigating the interaction of a model SCR with the seabed. Applied Ocean Research. 2010

[38] Jin J, Audibert JM, Kan WC. Practical design process for flowlines with lateral buckling. In: Proceedings of the ASME 2010 29th International Conference on Ocean, Offshore and Arctic Engineering, OMAE2010-20478. Shanghai, China: ASME; 2010

[39] Kimiaei M, Randolph M, Ting I. A parametric study on effects of environmental loadings on fatigue life of steel catenary risers (using a nonlinear cyclic riser-soil interaction model). In: Proceedings of the ASME 2010 29th International Conference on Ocean, Offshore and Arctic Engineering; Shanghai, China. ASME; 2010

[40] Nakhaee A, and Zhang J. Effects of the interaction with the seafloor on the fatigue life of a SCR. In: Proceedings of the 18th (2008) International Offshore and Polar Engineering Conference; 2008

[41] Nakhaee A, Zhang J. Trenching effects on dynamic behavior of a steel catenary riser. Ocean Engineering. 2010;**37**:277-288

[42] Aubeny CPA, Biscontin G. Seafloor-riser interaction model. International Journal of Geomechanics. 2009;**9**(3)

[43] Oliphant J, Maconochie A, White D, Bolton M. Trench interaction forces during lateral SCR movement in deepwater clays. In: Offshore Technology Conference; Houston, Texas, USA. 2009

[44] Clukey E, Young A, Dobias J, Garmon S. Soil response and stiffness laboratory measurements of SCR pipe/soil interaction. In: Offshore Technology Conference, OTC 19303, Houston, Texas. 2008

[45] Sen TK. Probability of fatigue failure in steel catenary riser in deep water. Journal of Engineering Mechnics. 2006

[46] Sen TK, Brown K. Root fatigue in deep water steel catenary risers—A probabilistic approach for assessment of risk. In: Offshore Technology Conference; 5-8 May 2008; Houston, Texas, USA

[47] Xia J, Das PK, Karunakaran D. A parametric design study for a semi/SCR system in northern north sea. Ocean Engineering. 2008;**35**:1686-1699

[48] Akpan U, Koko T, Rushton P, Tavassoli A, Else M. Probabilistic fatigue reliability of large diameter steel catenary risers (SCR) for ultra-deepwater operations. In: Proceedings of the 26th International Conference on Offshore Mechanics and Arctic Engineering. California, USA: ASME; 2007

[49] Giertsen E, Verley R, Schroder K. Carisima: A catenary riser/soil interaction model for global riser analysis. In: 23rd International Conference on Offshore Mechanics and Arctic Engineering, OMAE2004-51345; Vancouver, British Columbia, Canada. 2004

[50] Langner CG. Fatigue life improvment of steel catenary risers due to self-trenching at the touchdown point. In: Offshore Technology Conference, OTC 15104, Houston, Texas, USA; 2003

[51] Bridge C, Willis N. Steel catenary risers—Results and conclusions from large scale simulations of seabed interactions. In: International Conference on Deep Offshore Technology. New Orleans, USA: PennWell; 2002

[52] Willis NRT, and West PTJ. Interaction between deepwater catenary risers and a soft seabed: Large scale sea trials. In: Offshore Technology Conference; Houston, Texas OTC 13113. 2001

[53] Pesce CP, Aranha JAP, Martins CA. The soil rigidity effect in the touchdown boundary-layer of a catenary riser: Static problem. In: Proceedings of the 8th International Offshore and Polar Engineering Conference; 1998; Monterial, Canada

[54] Hale JR, Morris DV, Yen TS, Dunlap WA. Modelling pipeline behaviour on clay soils during storms. In: Offshore Technology Conference, OTC 7019; May 4-7 1992; Houston, Texas

[55] Grealish F, Kavanagh K, Connaire A, Batty P. Advanced nonlinear analysis methodologies for SCRs. In: Offshore Technology Conferecne, OTC 18922, Houston, Texas, USA. 2007

Chapter 4

Local Scour around a Monopile Foundation for Offshore Wind Turbines and Scour Effects on Structural Responses

Wen-Gang Qi and Fu-Ping Gao

Abstract

Monopile is the most commonly used foundation type for offshore wind turbines. The local scour at a monopile foundation generated by the incoming shear flow has significant influence on both quasi-static lateral responses and dynamic responses of the monopile. This chapter focuses particularly on characterizing the local scour in both spatial and temporal scales and revealing the scour mechanisms associated with the flow field around a monopile. The predicting methods for the equilibrium scour depth and the time scale of scour are detailed under various representative flow conditions in the marine environment. The scale effect while extrapolating the results of model tests to prototype conditions is highlighted. The local scour imposes significant influence not only on the deformation and stiffness of the monopile foundation, but also on the natural frequency and fatigue life of the structure system. Monopiles with diameters up to 10 m have become a feasible option as the industry is currently advancing into deeper waters. More meticulous considerations for monopile design associated with the scour depth prediction and evaluation of scour effects are still in need to efficiently minimize the cost while remaining safety simultaneously.

Keywords: local scour, monopile, combined waves and currents, horseshoe vortex, time scale, backfilling, scale effect, pile-soil interaction, natural frequency

1. Introduction

Offshore wind power is developing into a major energy source globally due to the abundant reserves, high-energy density, and fewer civil complaints compared to onshore wind power. The global installed capacity reached more than 14,300 MW at the end of 2016 [1], and an increased potential for growth in future was estimated [2, 3]. To achieve high energy harvesting efficiency, the industry is currently advancing into deeper waters accompanied by upscaling the offshore wind turbines (OWTs) from 5 to 8 MW, 10 MW and then 12 MW [4] (see **Figure 1**). The steady upward trends in both water depth and size of OWTs have led to consequential growth in the loading, further aggravating the deformation of the foundations and potentially jeopardizing the operation of the OWTs.

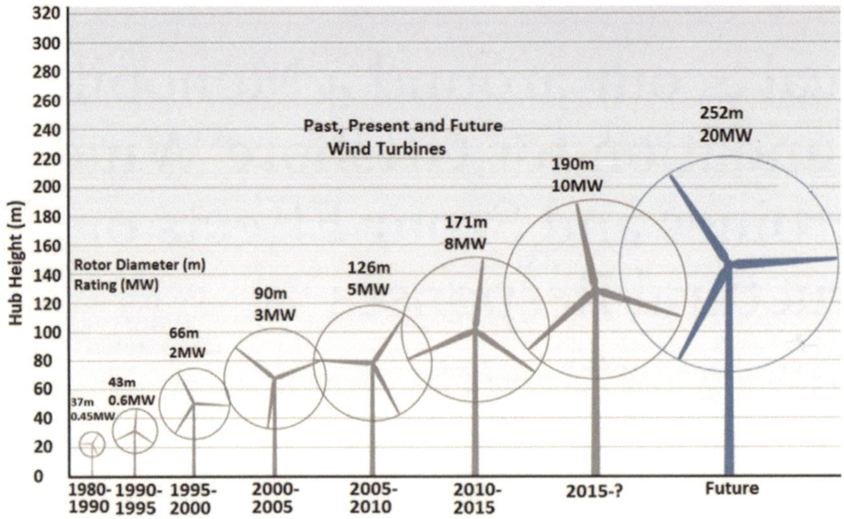

Figure 1.
Growth in size and power of horizontal axis wind turbine [4].

As the most commonly adopted foundation for OWTs, the monopile had been used as the foundation of 2653 OWTs by the end of 2016, which accounts for over 80% [5], due to their ease of design and manufacture in contrast to other foundation types. A typical monopile is composed of a hollow steel cylinder with a diameter typically spanning 4 to 8 m and a slenderness ratio (driving depth over diameter) less than 10 [5]. A special aspect to consider with monopile foundations for OWTs is that the vertical load mainly comprised of gravity load is relatively small compared with other offshore structures like oil and gas platforms. The horizontal loads generated by wind and wave actions, which are of similar magnitude to the gravity loads, are therefore of particular importance for the design [6]. A typical laterally loaded offshore monopile foundation generally behaves rigidly due to their large bending stiffness and low slenderness ratio [7]. The serviceability requirements including lateral deformation and stiffness rather than ultimate lateral resistance tend to govern design due to strict rotational tolerance specifications for monopile foundations [8].

Under the current and/or wave loadings, severe local scour can be generated around a monopile situated in cohesionless sediments. The maximum scour depth is generally predicted to be proportional to the pile diameter, and scour depths of over 1.5 times pile diameter were observed in many practical installations [9]. Considering the lower slenderness ratio compared with more conventional offshore pile for oil and gas industries, the maximum scour depth at a monopile may account for over 25% of the driving length [10]. This scour-induced driving length reduction has significant influence on both quasi-static lateral responses (e.g., ultimate lateral resistance and deformation stiffness) and dynamic responses (e.g. natural frequency) of monopile foundations. Therefore, proper assessments of local scour and its influence on structural responses are essential to the design of monopile foundation for OWTs.

In view of the complexity of local scour phenomenon around a monopile and scour effects on manifold monopile responses, this chapter aims to systematically summarize the predicting methods involving the maximum scour depth and the time scale under various flow conditions. The controlling mechanisms of local scour around a monopile are also revisited. Subsequently, we examine existing research on the effects of local scour on the responses of the monopile foundation.

2. Local scour around a monopile

2.1 General description and scour mechanisms

In the river hydraulics, the local scour at the bridge piers has been widely recognized to be one of the main causes for the structure failure [11]. Hence, the phenomenon of scouring at bridge piers has ever been studied extensively over the past half century from various aspects, including local scour mechanisms associated with flow characteristics, prediction of the maximum scour depth, etc. Comprehensive descriptions of scour around a pile or bridge pier in steady currents have been given in Refs. [12, 13].

For a pile foundation installed in the marine environment, several phenomena with regard to the flow pattern would arise including a horseshoe vortex (HSV, or sometimes necklace vortex due to its shape) in front of the pile, wave vortices convected through the pile's lee-side, and the contraction streamlines at the side edges of the pile. These changes induce an increased sediment transport capacity. If the scouring potential created by the large-scale eddy structures is strong enough to overcome the particles' resistance to motion, local scour will be initiated around the pile (see **Figure 2**). As scour hole develops, the strength of the large-scale eddy structures will gradually be weakened, thereby reducing the transport rate from the local scour region. Equilibrium is reestablished if the sediment inflow of the scour hole is equal to the outflow for live-bed scour, or the shear stress caused by vortices is equal to the critical shear stress of incipient sediment motion at the bottom of the scour hole for clear-water scour. Two major issues concerning the local scouring process, namely the equilibrium scour depth and the time scale, usually need to be quantitatively characterized and are detailed in Sections 2.2 and 2.3, respectively. We herein focus on reexamining the general scouring mechanisms, which mainly involve characterization of the flow pattern under different incoming flow conditions.

A surface-mounted cylinder induces the boundary layer separation from the upstream bed in the incoming flow, resulting in a HSV system at the upstream junction corner of the cylinder. The existing test observations indicated that the HSV (including the down-flow) upstream of a cylinder is the most evident and significant contributor to the scour process in currents alone. Much work has been done for the HSV at a cylinder in steady currents. Many researchers investigated the HSV at a flatbed for the laminar case [14–16]. However, most of the practical HSVs occur with turbulent approach boundary layers [17]. Baker [18] used smoke- and oil-flow visualization to observe the turbulent HSV system. The variation in the

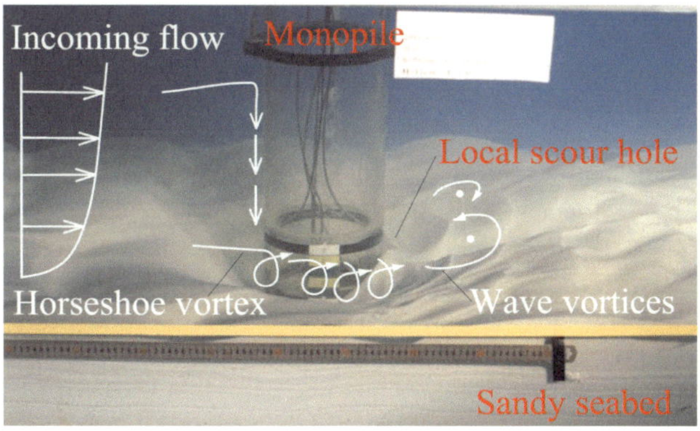

Figure 2.
General flow patterns involving local scour around a monopile.

position of the primary horseshoe vortex with the flow parameters was presented graphically. Dargahi [19] tried to examine the coupling between the flow field and scouring around a circular pile. The consequences of an unsteady vortex pattern were indicated. With the development of the scour hole, the vortex rapidly grows in size and strength as additional fluid attains a downward component and the strength of the down flow increases (see [20]). This conclusion was also verified by the subsequent flow measurements [21, 22].

Kirkil et al. [23] investigated the influence of large-scale coherent structures and macroturbulence events on the scour hole development at a cylinder. The results indicated that the pressure fluctuations and turbulent kinetic energy levels inside the primary necklace vortex were much larger than those associated with the turbulence inside the boundary layer flow approaching the cylinder. Strong ejections and uplift events were observed in the region just behind the cylinder. Recently, the evolution of the HSV in a scour hole around a circular pile was studied by Guan et al. [24] using 2D Particle Image Velocimetry (PIV) technique. The results indicated that the horseshoe vortex system evolves from one initial small vortex to three vortices along the upstream slope of the developing scour hole. Both the strength and size of the main vortex increase with increasing scour hole size. Most existing studies regarding the HSV at a scour hole in steady currents mainly focus on describing the evolution of the HSV itself rather than revealing the time-dependent coupling between the HSV and the scour hole. There is still a great need to improve the understanding of this complex coupling process (see [25]).

In contrast to the substantial amount of accumulated knowledge for HSV in steady currents, relatively fewer studies (e.g., [26]) can be found regarding the flow patterns around a cylinder under waves, although the two-dimensional vortex flow behind a free cylinder subject to an oscillatory flow is well understood (e.g., [27]). Sumer et al. [28] concluded that the scour around a pile in waves is governed by the action of both lee-wake vortices and the horseshoe vortex. Their observations showed that the lee-wake vortices shed from the cylinder act like cyclones, sweeping the sediment grains into its core region and then lifting the grains upward via the updraft in the vortices. The extension and circulation of the lee-wake vortices are governed primarily by the Keulegan Carpenter (*KC*) number [27]. The *KC* number is defined as

$$KC = U_{\text{wm}} T / D \tag{1}$$

where U_{wm} is the maximum velocity of the undisturbed wave-induced oscillatory flow at the sea bottom above the wave boundary layer, D is the pile diameter, and T is the wave period. In contrast to the controlling role of the lee-wake vortices in eroding the sediment grains around a cylinder, the HSV would be less significant because the wave boundary layer is usually thin (see [29, 30]), especially for relatively small *KC* numbers [26]. Regarding monopile foundations for OWTs, the corresponding value of *KC* number is usually low due to the large diameter [31]. Therefore, the effect of the HSV might be insignificant for the scour development around a monopile under waves.

In general, the controlling mechanism of the scour process around pile foundation in wave-alone or in current-alone condition has been well established. Yet for the condition of combined waves and currents, which is a common scenario in marine environments, the physical mechanism for the scouring phenomenon is less understood. When the waves and currents coexist, the sediment is normally picked up by the waves due to its higher capacity of lifting sands and transported by the currents due to its higher capacity of carrying sands. Nevertheless, the effect of waves and currents' coexistence is more than just a superimposition of their capacities of initiating and carrying sediment, due to a nonlinear interaction between waves and currents within and outside the bottom boundary layer [32–34]. The only experimental investigation so far on the flow pattern around a vertical pile in a combined current-wave flow was made by Sumer et al. [26]. Their results showed that introducing the following currents in the waves increases the size and lifespan of the horseshoe vortex and lowers the critical KC number for the threshold of horseshoe vortex. The horseshoe vortex exists for smaller KC with increasing the current-to-wave velocity ratio. This result is related to the increase in the adverse pressure gradient in front of the pile caused by the superimposed currents. For small values of KC (such as $KC \approx O(10)$), the presence of the horseshoe vortex is quite limited in both space and time domain, and its influence on the scour is much less than that in steady currents ($KC \approx \infty$). That is, with decreasing KC, the effect of horseshoe vortex decreases and that of the vortex shedding increases accordingly.

2.2 Equilibrium scour depth

The equilibrium scour depth prediction is a key concern in the geotechnical design for the coastal and offshore foundations. On the basis of laboratory flume tests and field data, many empirical equations have been proposed with respect to the equilibrium scour depth around a cylinder over the past few decades.

For the bridge scour in steady currents, Melville and Sutherland equation [35] and the Colorado State University (CSU) equation [36] are two of the most commonly used scour equations. They give somewhat conservative estimates of scour depth, the former perhaps more so than the latter [37]. Other well-known scour equations were reviewed by Deng and Cai [38]. The accuracy of several most commonly adopted scour equations were evaluated and compared by Johnson [37], Mueller [39] and Sheppard et al. [40], using a large set of field data.

Unlike the pier-scour in bridge waterways, the local scour at offshore monopile foundations should consider the action of waves. For scour depth around pile exposed to waves alone, the number of prediction equation is much fewer compared with that under steady currents. An empirical expression for scour depth at a circular slender pile exposed to regular waves was established [28]:

$$S/D \approx 1.3[1 - \exp(-0.03(KC - 6))] \text{ for } KC \geq 6 \tag{2}$$

where S is the equilibrium scour depth at the pile. Note that Eq. (2) is valid for the live-bed conditions. Sumer et al. [41] further investigated the effect of the soil density on the scour depth at a pile in waves. The scour depth was significantly increased when the bed soil was changed from medium dense sand to dense sand. Dey et al. [42] presented the results of equilibrium scour depths at circular piles in clay and sand-clay mixed beds under waves alone. The variation of equilibrium scour depth with KC for different soil conditions follows an exponential law, as was given in [28], having different coefficients and exponents. Recently, Zanke et al. [43] proposed the following unifying equation for the prediction of equilibrium scour depth around a pile under the action of waves, tidal or steady currents.

$$S/D = 2.5(1–0.5\,U_{cr}/U)\,x_{rel} \tag{3}$$

where $x_{rel} = x_{eff}/(1 + x_{eff})$, $x_{eff} = 0.03(1 - 0.35U_{cr}/U)\ (KC\text{-}6)$, U is the mean velocity in case of steady currents, and U_{cr} is the critical velocity for initiation of sediment motion.

For shallow-water subsea locations, waves often coexist with currents, propagating either following or opposing the currents. This brings the soil response around marine structures more complicated than that due to waves or currents alone. Nevertheless, the related research is relatively few. Sumer and Fredsøe [44] conducted a series of tests of irregular waves propagating either along or perpendicular to the current, indicating that the scour depth for combined waves and currents is influenced by the KC number and the current-to-wave velocity ratio. The current-to-wave velocity ratio is defined as $U_{cw} = U_c/(U_c + U_{wm})$, in which U_c represents the undisturbed near-bed current velocity component of the combined flow, and U_{wm} is the maximum velocity of the undisturbed wave-induced oscillatory flow at the sea bottom above the wave boundary layer. Sumer and Fredsøe [45] carried out re-analyses of scour data in [44] and derived the following empirical expression for the scour depth under combined waves and current:

$$S/D = S_c/D\{1 - \exp[-A(KC - B)]\} \text{ for } KC \geq B \quad (4)$$

in which S_c is the scour depth for the current-only case, and the parameters A and B are given as $A = 0.03 + 0.75U_{cw}^{2.6}$ and $B = 6\exp(-4.7U_{cw})$. The validity range of Eq. (4) is limited to $4 < KC < 25$. The value of S_c/D can be expressed with a mean value $S_c/D = 1.3$ and a standard deviation $\sigma_{S/D} = 0.7$. The maximum scour depth for the live-bed scour regime is suggested to take as $S_c/D = 1.3 + \sigma_{S/D} = 2.0$ or $S_c/D = 1.3 + 2\sigma_{S/D} = 2.7$ for design purpose. The large variation range of S_c/D in Eq. (4) could produce a great arbitrariness and uncertainty in the practical engineering application. Moreover, the accuracy of Eq. (4) is not very much optimistic [46]. Rudolph and Bos [47] carried out model tests on scour around a monopile under combined waves and currents with oblique directions, focusing on the range $1 < KC < 10$. They proposed an improved scour depth prediction formula based on Sumer and Fredsøe [45] by analyzing their new data and preceding data.

Recently, Qi and Gao [31, 48] conducted an experimental investigation of the local scour development under combined waves and currents at a monopile. The influence of the wave propagating direction relative to currents is obvious. The experimental results indicated that the wave-current interaction had a substantial effect on the time-development of local scour and the resulting equilibrium scour depth. An average-velocity-based Froude number (Fr_a) was proposed to correlate with the equilibrium scour depth (S/D) at offshore monopile foundation in combined waves and currents [48]. The definition of Fr_a is

$$Fr_a = U_a/\sqrt{gD} \quad (5)$$

in which U_a $\left(= \frac{1}{T/4}\int_0^{T/4} (U_c + U_{wm}\sin(2\pi t/T))dt = U_c + \frac{2}{\pi}U_{wm}\right)$ is the average water particle velocity during one-quarter cycle of oscillation under combined waves and currents, when the oscillatory motion and the currents are in the same direction; t is the time and g denotes the gravitational acceleration. The equilibrium scour depths (live-bed scour regime) under combined waves and currents obtained from the experiments were plotted as a function of Fr_a (see **Figure 3a**), along with those of in [44, 49]. Following empirical expression for the correlation between S/D and Fr_a was given for predicting equilibrium scour depth under live-bed conditions.

$$\lg(S/D) = -0.8\exp(0.14/Fr_a) + 1.11 \; (0.1 < Fr_a < 1.1, 0.4 < KC < 26) \quad (6)$$

Eq. (6) indicates that the scour depth approaches its mathematical asymptotic value ($S/D \approx 2.0$) as Fr_a increases (e.g., larger than 1.0). Note that this asymptotic value is in the range of a typical scour depth prediction ($S/D \approx 1.7–2.4$) by previous local scour equations around a pile for current-alone conditions. The prediction

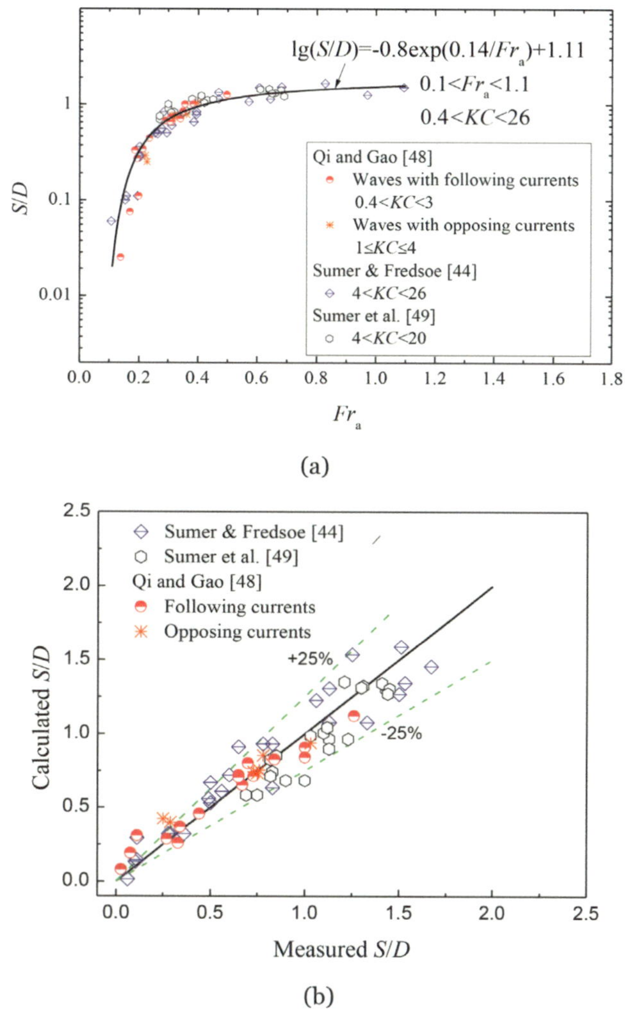

Figure 3.
(a) S/D vs. Fr_a for combined waves and currents; (b) comparison of measured and predicted scour depths [48].

using the empirical correlation generally distributes in the range of 25% error lines (see **Figure 3b**), which is superior to the existing formula.

As aforementioned, most existing equations for predicting the equilibrium scour depth around a cylinder are based predominantly on scaled laboratory flume data. Therefore, the inherent scaling issue is worthy of clarifying before extrapolating the results of small scale scour experiments to prototype [50]. The scale effect is essentially attributable to the use of three independent length scales in local scour experiments: cylinder diameter, sediment size, and water depth [51]. The three length scales lead to several controlling similarity parameters that inevitably cannot be satisfied concurrently. Two aspects are specifically drawing attention involving the scale effect: one associated with inadequate similitude of large scale turbulence generated around a cylinder during scour tests and the other with the distortion of sediment size scaling.

The similitude of large scale turbulence requires preservation of flow patterns so that pressure heads along flow path scale directly with the geometric scale relating the small cylinder in a flume test to a prototype pile. The Froude number defined with incoming flow velocity and pile diameter is proved useful for reflecting the similitude associated with flow-field [51, 52]. It was concluded that the flume-based equations for predicting equilibrium scour depth at a cylinder may not adequately account for the similitude of large scale turbulence and thus commonly lead to over-conservative estimations in prototype applications [51].

Most flume experiments were forced to use particles of comparable size to full-scale particles due to requirements of both the similitude associated with particle mobility and the natural size limit to which bed particles become cohesive. As such, the similarity of sediment size with respect to pile diameter cannot be maintained. According to Melville and Chiew [53], D/d_{50} barely has effect on the scour development for $D/d_{50} > 50$, with d_{50} as the median grain size of the sediment. However, some more recent research indicates a decrease in S/D at very large values of D/d_{50} [54, 55]. Consequently, a prototype scour depth prediction based on the flume scour data is greater than that in the field, that is, a relatively conservative or safe prediction. A comparison between the maximum calculated scour data from different formulations used in standard and technical recommendations for the design of offshore wind farms and available scour data in European wind farms found in technical reports was once conducted by Matutano et al. [9]. The observed maximum scour depths were largely less than estimated in most offshore wind farms studied, which is consistent with the similarity analysis.

It should be noted that various offshore standards like DNV·GL [8] recommend Eq. (2) proposed in [26] to estimate the scour depth around monopile foundations for the OWTs under combined waves and currents. This could be inappropriate and potentially lead to great inaccuracies [56].

2.3 Time scale

The time scale of scour process (T_0) is a certain amount of elapsed time for a substantial amount of scour to develop [45], which is essential for predicting the scour depth at any instant in a certain wave and/or current condition. In the case of live-bed scour, the equilibrium scour depth is achieved faster than the clear-water case, generating a smaller time scale. The time scale of scour process can be defined in several ways according to different mathematical functions for approximating the scour development toward the equilibrium stage [57]. For example, Briaud et al. [58] proposed a hyperbolic function $S_t = S_e\left(\frac{t}{t+T_0}\right)$, where S_e is the equilibrium scour depth. Alternatively, a more commonly used definition was proposed in [59]: $S_t = S_e\left(1 - \exp\left(-\frac{t}{T_0}\right)\right)$. The value of the time scale can be obtained by fitting the measured scour depth development curve with these functions of S_t. The normalized time scale is usually defined as

$$T^* = T_0 \frac{\left(g(s-1)d_{50}^3\right)^{\frac{1}{2}}}{D^2} \quad (7)$$

where s is the specific gravity of sand grains.

For steady currents, the normalized time scale is associated with the ratio of boundary layer thickness to pile diameter (δ/D) and the Shields parameter (θ), using following empirical expression for live-bed scour regime [59].

$$T^* = \frac{1}{2000}\frac{\delta}{D}\theta^{-2.2} \quad (8)$$

The Shields parameter (θ) is defined as

$$\theta = \frac{U_f^2}{g(\rho_s/\rho_w - 1)d_{50}} \tag{9}$$

in which U_f is the maximum value of the undisturbed friction velocity, ρ_s is the sediment grain density, and ρ_w is the water density. The detailed calculation procedure of θ is given by Soulsby [60].

In the case of waves, the time scale of the scour process around a cylinder is governed by two parameters, namely the KC number and the Shields parameter. The empirical expression of T^* based on KC and θ for live-bed scour regime can be represented by [59].

$$T^* = 10^{-6}\left(\frac{KC}{\theta}\right)^3 \tag{10}$$

Under combined waves and currents, Petersen et al. [61] concluded that the time scale of scour varies with the KC number, the Shields parameter associated with the wave component of the flow (θ_w), and U_{cw}. The time scale of scour increases significantly when even a slight current is superimposing on a wave. On the basis of the data of Refs. [61, 62] and their own, Chen and Li [63] proposed a piecewise fitting formula for the time scale of scour under combined waves and currents (see **Figure 4**)

$$T_0{}^* = \begin{cases} 8.67 \times 10^{-5}\theta_w^{-5.15}U_{cw}^{28.1(0.152-\theta)} & \text{for } 0.3 < U_{cw} < 0.44 \\ 1.35 \times 10^{-4}\theta_w^{-2.2}U_{cw}^{-6.8} & \text{for } 0.44 \le U_{cw} \le 0.7 \end{cases} \tag{11}$$

2.4 Backfilling process

Backfilling process is the deposition of sediment into a scour hole previously generated, when the pile is exposed to a changing flow environment (the flow

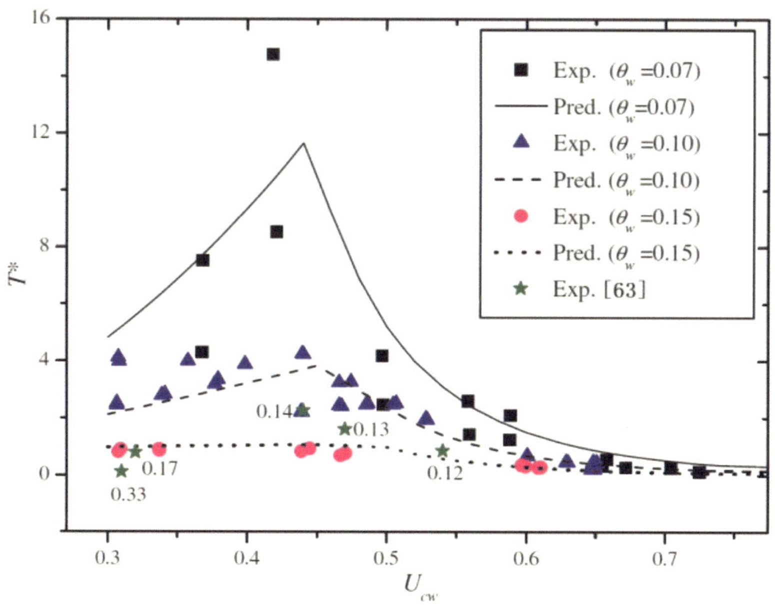

Figure 4.
Comparison of predictions using Eq. (11) and calculated results of normalized time scale based on observed data under combined wave-current conditions [63].

environment changing from, e.g., a steady current to a wave, from a steady current to combined waves and currents, or from a wave to a smaller wave). For the monopile foundations of the OWTs, the flow climate changes continuously, hence leading to scour and backfilling in an alternating fashion. The backfilling process needs to be quantitatively estimated to establish models for predicting the time history of scour and backfilling for large periods of time (see [64]).

Hartvig et al. [65] experimentally studied the scour and backfilling around a pile subjected to alternating sequences of currents with and without superimposed irregular waves. The results indicated that initially current-scoured beds, when subjected to waves, are backfilled about 10 times slower than the scoured ones. Sumer et al. [49] later presented a systematic investigation of the scour holes backfilling under waves or combined waves and currents around a pile. The results showed that the scour depth corresponding to the equilibrium state of backfilling is the same as that corresponding to the equilibrium state of scour around the pile for the same wave or combined waves and currents climate. That is, the final equilibrium scour is uninfluenced by the loading history of the flow but only dependent on the flow condition at the last stage if the last loading stage is sufficiently long. The time scale of backfilling is much larger than that of scour when the KC number associated with the backfilling is $KC < O(10)$ (typical wind farm application). Various empirical equations involving the time scale were further proposed for different flow environment changes.

Backfilling can reduce the scale of the scour hole, which is beneficial to the bearing behavior of a monopile unprotected against scour. Sørensen et al. [66] assessed the relative density of the backfilled soil around a model monopile of 0.55 m diameter based on the test in a large wave channel. The relative density of the backfill material was found to be rather large, varying between 65 and 80%. Note that these values were obtained from only one set of laboratory tests, and lower densities may exist. The backfilled soil was found to increase the soil support to a monopile foundation significantly. Therefore, an optimized design with respect to both the fatigue limit state and the ultimate limit state can be achieved by considering the time variation of the scour depth [67].

3. Effects of local scour on the lateral response of a monopile

3.1 Lateral bearing behavior

The lateral response of a pile is commonly evaluated using a load transfer approach, with "p-y" curves quantifying the nonlinear interaction between the pile and the surrounding soil. The p-y curves idealize the soil as a series of independent springs distributed along the pile length, with each spring describing the nonlinear relationship between the lateral soil resistance (p) and the lateral deflection of the pile (y). Various formulations for p-y curves have been developed for sands, for example Refs. [68–71]. The results in Refs. [68, 69] form the basis of the widely used API method [72].

Scour around submarine structures leads to a reduction in the effective stresses and hence strength and stiffness of the soil at a given depth around a pile. As scour depth is predicted to scale with the pile diameter, scouring can impact greatly on the overall response of large-diameter monopiles due to the low slenderness ratio (typically between about 3 and 10). As shown in **Figure 5**, scour reduces the pile embedment depth from L to L' and increases the load eccentricity (cantilever length) from e to e', changing the profiles of lateral soil resistance p, inducing much larger deformations of the pile and reducing the foundations stiffness.

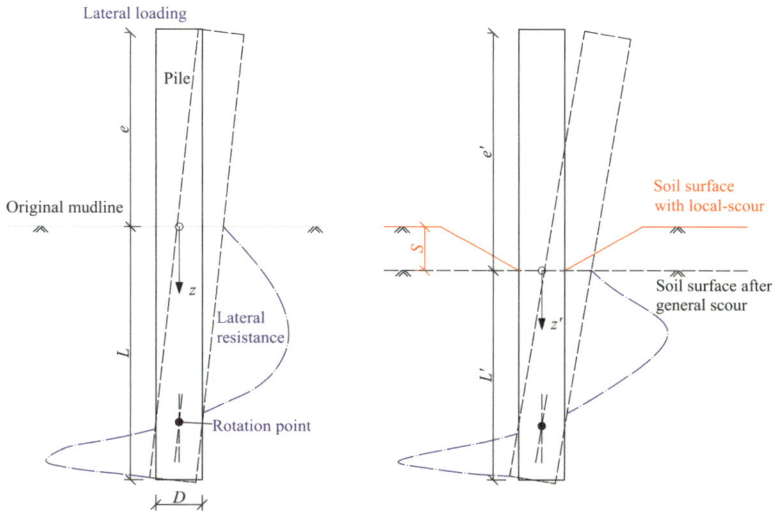

Figure 5.
Illustration of scour effect on the lateral response of a shallowly embedded pile [73].

Scour can cause the *p-y* response at a given absolute depth z (measured from the original soil surface or "mudline") to soften, reducing both initial stiffness and the limiting resistance. In design codes such as DNV·GL [8], local scour is often considered to cause complete loss of soil resistance down to the depth of scour below the original seabed while estimating the pile capacities. Although such a design practice is simple, it often leads to extreme conservative pile foundation design and subsequent unnecessary cost [73]. In recent years, the importance of scour hole dimensions for pile capacities has been increasingly recognized.

In an analytical study [74], general scour effects were incorporated by updating the parameters per unit length (p_{ult}) at a given depth to account for the overconsolidation induced by removal of a scour depth of sand. Lin et al. [75] and Yang et al. [76] further modified the ultimate lateral resistance due to changes in the shallow wedge-type failure mechanism to consider the effects of the local scour. Results indicated that the local scour hole would result in much higher lateral soil resistance for a given depth than for the general scour case, that is, complete removal of the soil surface layer. Tseng et al. [77] combined the approach of Sørensen [78] for modifying the initial modulus of subgrade reaction (k_{py}) and the approach of Lin et al. [75] for modifying p_{ult} to evaluate the scour effect on the deformation response of monopile foundations. According to the case analysis, when the *p-y* curve method suggested by the design guidelines for an offshore wind turbine was used to design the monopile foundation for an OWT, the foundation deformation was underestimated for a scour depth of less than pile diameter, and foundation stiffness was underestimated for a scour depth of greater than pile diameter.

A series of centrifuge tests were conducted by Qi et al. [73] to investigate the scour effects on *p-y* curves. A practical approach to incorporate effects of scour on the *p-y* curves was proposed by adopting an effective soil depth (z_e) in the determination of *p-y* curves. The effective soil depth was expressed by

$$\frac{z_e}{D} = \frac{z'}{D} + \tanh\left(f\frac{z'}{D}\right)\frac{S}{D} \qquad \begin{cases} f \approx 1.5 & \text{for local-scour} \\ f = 0 & \text{for general-scour} \end{cases} \tag{12}$$

where the effective soil depth z_e is a weighted-average of the soil depth relative to the original mudline z and the soil depth below the current scour base z' ($= z - S$,

S is the scour depth), f is an empirical parameter indicating the transition rate for the effective soil depth from $z_e = z'$ at the current mudline to $z_e \approx z$ as the soil depth increases. The recommended value of $f = 1.5$ for local-scour conditions was based on the tests with 30° slope angle of the scour hole. As the slope angle of the scour hole decreases, the effect of the sloping overburden soil above the level of the scour base reduces and the value of f should be reduced accordingly. This conclusion conforms to the results of a recent study by Lin and Wu [79].

Local scour not only exaggerates the deformation and reduces the stiffness of the monopile foundation, but also changes its bearing mechanism. Monopile foundations for the OWTs generally behave rigidly, rotating as a whole and developing a "toe-kick" phenomenon. **Figure 6** illustrates the mobilized soil resistance components of a laterally loaded short pile. Significant base shear and axial resistance will be mobilized at the pile tip if the slenderness ratio gets sufficiently small. Moreover, vertical side resistances emerge around the pile shaft due to the mobilized horizontal displacement of the pile tip and rotation of the pile. Numerical investigations by Bekken [80] and Byrne et al. [81] have confirmed the significant contributions due to the horizontal pile tip displacement and rotation of the pile on the resistance of a monopile.

To investigate the scour effect on the behavior of a laterally loaded monopile, a three-dimensional finite element model was proposed and verified by Qi and Gao [83]. The results indicated that the scour could induce a significant transition of pile behavior for a typical monopile foundation of the OWTs. **Figure 7(a)** shows the variation of y_p/y_n with E_sL^4/E_pI_p, in which y_n denotes the lateral pile displacement at the loading position obtained from the 3D finite element results, and y_p denotes the lateral pile displacement according to the conventional p-y method. The non-dimensional expression of E_sL^4/E_pI_p is aimed to reflect the rigidity of the pile foundation [7, 84], where I_p denotes the second moment of area of a pile and E_pI_p

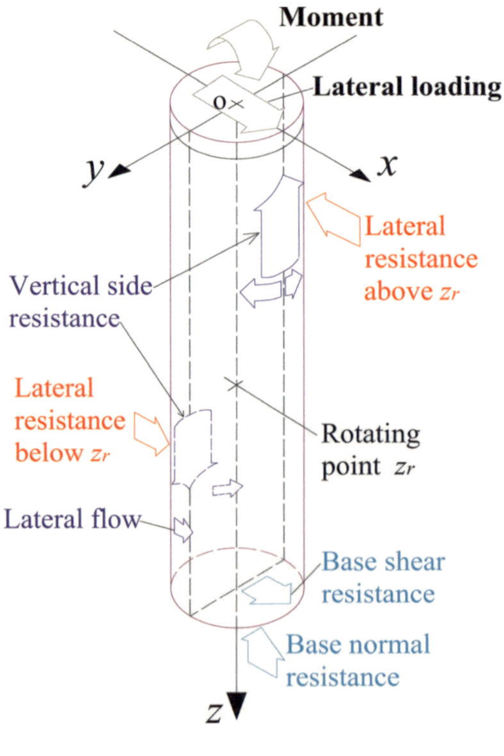

Figure 6.
Resistance components of a laterally loaded monopile foundation [82].

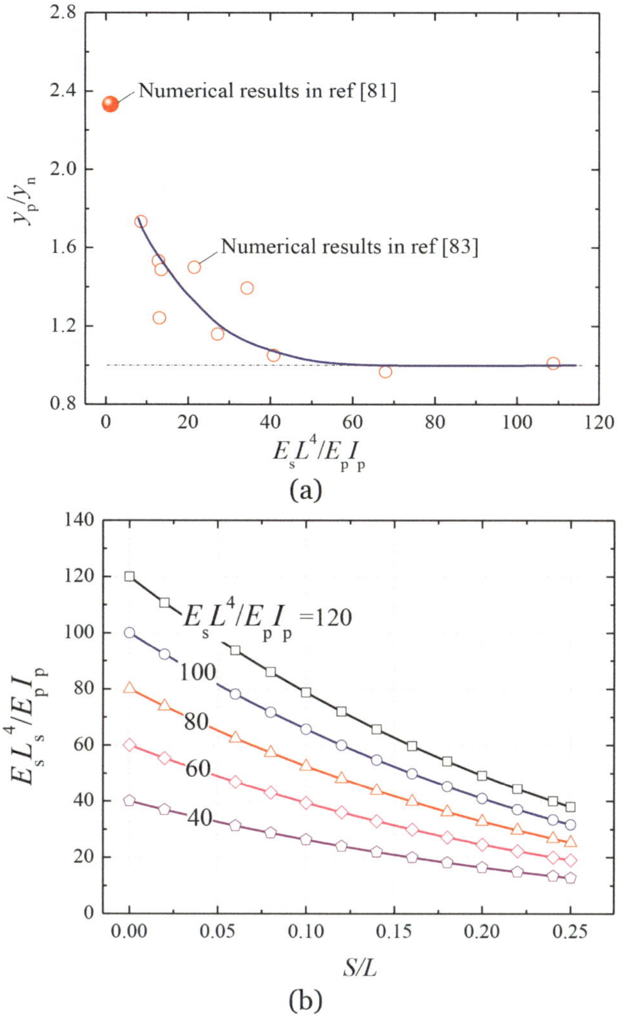

Figure 7.
(a) Variation of y_p/y_n with E_sL^4/E_pI_p; (b) effect of scour depth normalized with pile embedment depth S/L on $E_sL_s^4/E_pI_p$ [83].

represents the bending stiffness of the pile. It is seen that for an approximate range of $E_sL^4/E_pI_p > 50$, the value of y_p/y_n is generally equal to 1.0, while the value of y_p/y_n obviously deviates from the value of $y_p/y_n = 1.0$ for $E_sL^4/E_pI_p < 50$. The deviation of y_p/y_n implies that due to the scour-induced increase of pile rigidity (see **Figure 7(b)**), the pile's tip and shaft resistance could become a significant component of the whole soil resistance. The parameter L_s in **Figure 7(b)** denotes the pile embedment depth after scour. The traditional *p-y* approach only takes account of the lateral soil resistance and thus the scour would render the traditional *p-y* approach inapplicable for post-scour conditions.

There are relatively fewer studies concerning the scour effect on the pile behavior under cyclic loading compared with aforementioned studies on the monotonic pile behavior [85]. Achmus et al. [86] numerically quantified the scour-induced increase of pile deformation by adopting a degradation stiffness model to account for cyclic loading. The results indicated an evident accumulation of lateral deformation under cyclic loading. Up to now, no systematic investigations into the influence of scour on the cyclic pile behavior can be found in the literatures. This issue needs to be further studied. Moreover, local scour and backfilling occurs around a monopile foundation in an alternating fashion due to the continuously changing flow climate in marine conditions. This variation of scour depth should also be included while evaluating the long-term cyclic pile behavior.

3.2 Natural frequency and fatigue life (FL)

Compared to other types of OWT foundations such as gravity base, tripod, and suction bucket, the monopile is a dynamically sensitive structure due to its flexibility. The primary excitation forces arise from the wind, waves, and the rotor excitations. The fundamental frequency f_0 (the first tower bending frequency) of the structure is the most important controlling parameter for its dynamic response.

The OWTs are usually designed as soft-stiff structures in order to avoid resonances, meaning that f_0 is above the rotational frequency of the rotor f_{1P} but below the blade-passing frequency f_{3P} (see **Figure 8**). The excitation frequencies related to environmental loads from wind and waves are typically below the 1P frequency for OWTs with relatively small installed capacities, for example, smaller than 3.6 MW [87, 88]. Nevertheless, for a 6–8 MW modern OWT on a monopile, designed for the soft-stiff frequency range, f_0 would be around 0.20–0.23 Hz [89]. In this frequency range, the fundamental mode of the structure would resonate with the waves at low wind speeds and be exposed to high spectral energy of waves during higher wind speeds. Due to the narrow soft-stiff frequency range for the OWTs on monopile foundations, even a small natural frequency shift from the initial design value may finally lead to resonance and the resulting stress in the materials of the monopile will be higher than in the quasi-static case. This high stress would potentially reduce the FL of the OWTs significantly.

Scour around the monopile leads to significant reduction of f_0. This would make f_0 align with the frequency of the passing sea waves or the rotational frequency of the rotor (f_{1p}) under certain circumstances, exposing structures to the threat of resonances. Sørensen and Ibsen [67] calculated the undamped natural frequencies of an existing offshore wind farm at HornsRev1 by means of the Winkler model approach. The stiffness derived from *p-y* curves given in the API code was adopted in their desk study, without accounting for the degradation of stiffness due to cyclic loading. The results revealed that the first natural frequency decreases with approximately 5% when the scour depth reached 1.3*D*. Prendergast et al. [90] conducted finite element numerical modeling to examine the scour effects on the natural frequency of a 3.6 MW OWT on a monopile in typical offshore sand deposits. The numerical model was validated by scale model tests and proved capable of tracking the scour-induced change in the natural frequency of the system. A reduction of up to 8.5% in natural frequency was observed in loose sand for

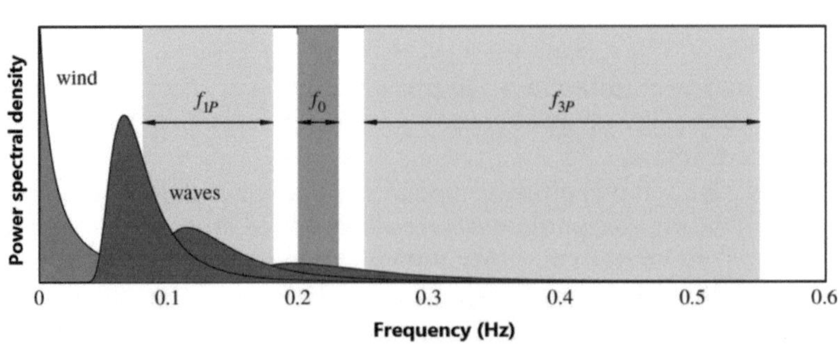

Figure 8.
Illustration of typical excitation ranges of a modern 6–8 MW OWT. Wave spectra represent different wind speeds. Not in scale [89].

S/D = 1.3. A comparison among cases with different soil stiffness profiles suggests that scour-induced resonance of an OWT on a monopile is most likely to occur in loose deposits. Prendergast et al. [91] recently improved the numerical model to further include the effect of spatial variability in soil properties derived from measured Cone Penetration Test (CPT) data. A stochastic ground model was developed to account for the uncertainty in soil-structure interaction stiffness for a given offshore site. The results indicated that the first natural system frequency reduces while the range of predicted system frequencies increases as the scour progresses. This means that the results of frequency-based SHM (structural health monitoring) technique for assessing scour magnitude-based solely on first natural frequency measurements becomes less certain for a deeper scour hole.

Several recent studies have focused on the scour effects associated with the FL of OWTs. Damgaard et al. [92] performed fully coupled aero-hydro-elastic simulations to evaluate the scour effects on the fatigue loads for parked OWTs. It was shown that a scour hole of 1.3D deep increases the fatigue damage equivalent moment with approximately 40%. Note that time-varying stiffness/damping properties of the soil due to long-term cyclic loading [93, 94] and scour depth variation due to backfilling are not accounted for. Rezaei et al. [95] presented a study of scour and backfilling effects on the FL of a 5 MW monopile supported OWT, using quasi-static, modal, and time-domain fatigue analyses. It was demonstrated that a scour depth of 1.5D could result in a 45% reduction of the FL of the wind turbine structure. Backfilling is beneficial to extend FL of the OWTs. Nevertheless, the relative density barely has effect on the FL of the OWTs.

4. Concluding remarks

Monopile is the most commonly adopted foundation type for OWTs. Local scour induced by waves and/or currents can jeopardize the function of the structure system to a great extent. In contrast to bridge scour which has been sufficiently studied over the past half century, an important feature of scour phenomenon around an offshore monopile stems from the various flow conditions in a marine environment, namely waves only, currents only, and combined waves and currents. In view of the complexity of incoming flow conditions, several crucial aspects related to local scour around a monopile foundation are reviewed, including the fundamental scour mechanisms, the maximum scour depth, the time scale, and the backfilling process. The inherent scaling issue arising from extrapolating the results of small scale scour experiments to prototype is clarified.

The driving depth of an offshore monopile can be significantly reduced by the local scour due to the low slenderness ratio, consequently imposing great influence on twofold structural responses. First, scour will cause the p-y response at a given absolute depth to soften, reducing both initial stiffness and the limiting resistance. Moreover, scour would render the traditional p-y approach inapplicable for post-scour conditions due to the change of bearing mechanism. Second, scour around the monopile leads to significant reduction of the fundamental frequency. This would make the fundamental frequency align with the frequency of the passing sea waves or the rotational frequency of the rotor under certain circumstances, exposing structures to the threat of resonances.

The current upward trends in both water depth and size of OWTs have led to consequential growth mainly in the horizontal loading. Monopiles with diameters up to 10 m have been proposed as a feasible option to support the OWTs with approximately 10 MW installed capacity. To efficiently minimize the cost while remaining safety, more meticulous considerations for monopile design are in need involving the scour depth prediction and evaluation of scour effects on structural responses. An alternative hybrid foundation system comprising a monopile and a bearing plate (or skirted footing) has been demonstrated to enhance the lateral bearing capacity compared with a monopile [96–99]. Nevertheless, the enhanced capacity of the hybrid foundation largely relies on the proper interaction between the bearing plate and the soils at the shallow zone of the seabed, where local scour and wave-induced pore pressure could occur [100–108]. The extent of local scour and scour effects on both the bearing capacity and natural frequency of a hybrid monopile need to be further investigated.

Acknowledgements

This work is financially supported by the National Natural Science Foundation of China (Grant Nos. 11602273; 11825205) and the Youth Innovation Promotion Association of the Chinese Academy of Sciences (Grant No. 2012014).

Author details

Wen-Gang Qi and Fu-Ping Gao*
Key Laboratory for Mechanics in Fluid Solid Coupling Systems, Institute of Mechanics, Chinese Academy of Sciences, Beijing, China

*Address all correspondence to: fpgao@imech.ac.cn

References

[1] GWEC. Global Wind Report 2016. 8th National Renewable Energy Forum. Ulaanbaatar, Mongolia: Global Wind Energy Council; 2017

[2] Perveen R, Kishor N, Mohanty SR. Off-shore wind farm development: present status and challenges. Renewable and Sustainable Energy Reviews. 2014;**29**:780-792. DOI: 10.1016/j.rser.2013.08.108

[3] Wu XN, Hu Y, Li Y, Yang J, et al. Foundations of offshore wind turbines: A review. Renewable and Sustainable Energy Reviews. 2019;**104**:379-393. DOI: 10.1016/j.rser.2019.01.012

[4] Igwemezie V, Mehmanparast A, Kolios A. Current trend in offshore wind energy sector and material requirements for fatigue resistance improvement in large wind turbine support structures—A review. Renewable and Sustainable Energy Reviews. 2019;**101**:181-196. DOI: 10.1016/j.rser.2018.11.002

[5] Negro V, López-Gutiérrez JS, Esteban MD, Alberdi P, Imaz M, José-María S. Monopiles in offshore wind: preliminary estimate of main dimensions. Ocean Engineering. 2017; **133**:253-261. DOI: 10.1016/j.oceaneng.2017.02.011

[6] Kuo YS, Achmus M, Abdel-Rahman K. Minimum embedded length of cyclic horizontally loaded monopiles. Journal of Geotechnical and Geoenvironmental Engineering. 2012;**138**(3):357-363. DOI: 10.1061/(ASCE)GT.1943-5606.0000602

[7] LeBlanc C, Houlsby GT, Byrne BW. Response of stiff piles in sand to long-term cyclic lateral loading. Géotechnique. 2010;**60**(2):79-90. DOI: 10.1680/geot.7.00196

[8] DNV·GL. DNV-OS-J101: Design of Offshore Wind Turbine Structures. 2013

[9] Matutano C, Negro V, JS L–G, et al. Scour prediction and scour protections in offshore wind farms. Renewable Energy. 2013;**57**/:358-365. DOI: 10.1016/j.renene.2013.01.048

[10] Sørensen SPH, Ibsen LB, Frigaard P. Experimental evaluation of backfill in scour holes around offshore monopoles. In: Frontiers in Offshore Geotechnics II. London, UK: Taylor & Francis Group; 2011. pp. 617-622

[11] Melville BW, Coleman SE. Bridge Scour. CO, USA: Water Resources Publications; 2000

[12] Hoffmans GJCM, Verheij HJ. Scour Manual. Rotterdam: A.A. Balkema; 1997

[13] Whitehouse R. Scour at Marine Structures: A Manual for Practical Applications. London: Thomas Telford; 1998

[14] Baker CJ. The laminar horseshoe vortex. Journal of Fluid Mechanics. 1979;**95**(2):347-367. DOI: 10.1017/S0022112079001506

[15] Seal CV, Smith CR, Rockwell D. Dynamics of the vorticity distribution in endwall junctions. AIAA Journal. 1997;**35**(6):1041-1047. DOI: 10.2514/2.192

[16] Bo H, Hua Z, Younis MY, Yan L, Raza MS. Experimental investigation on the transition of separation/attachment in steady laminar juncture flows. Experiments in Fluids. 2015;**56**(4):1-9. DOI: 10.1007/s00348-015-1943-5

[17] Simpson RL. Junction flows. Annual Review of Fluid Mechanics. 2001; **33**(33):415-443

[18] Baker CJ. The turbulent horseshoe vortex. Journal of Wind Engineering and Industrial Aerodynamics. 1980;**6**

(1–2):9-23. DOI: 10.1016/0167-6105(80)90018-5

[19] Dargahi B. Controlling mechanism of local scouring. Journal of Hydraulic Engineering. 1990;**116**(10):1197-1214. DOI: 10.1061/(ASCE)0733-9429(1990)116:10(1197)

[20] Breusers H, Nicollett G, Shen H. Local scour around cylindrical piers. Journal of Hydraulic Research. 1977;**15**: 211-252. DOI: 10.1080/00221687709499645

[21] Dey S, Raikar RV. Characteristics of horseshoe vortex in developing scour holes at piers. Journal of Hydraulic Engineering. 2007;**133**(4):399-413. DOI: 10.1061/(ASCE)0733-9429(2007)133:4(399)

[22] Unger J, Hager WH. Down-flow and horseshoe vortex characteristics of sediment embedded bridge piers. Experiments in Fluids. 2007;**42**(1):1-19. DOI: 10.1007/s00348-006-0209-7

[23] Kirkil G, Constantinescu SG, Ettema R. Coherent structures in the flow field around a circular cylinder with scour hole. Journal of Hydraulic Engineering. 2008;**134**(5):572-587. DOI: 10.1061/(ASCE)0733-9429(2008)134:5(572)

[24] Guan DW, Chiew YM, Wei MX. Characterization of horseshoe vortex in a developing scour hole at a cylindrical bridge pier. International Journal of Sediment Research. 2019;**34**(2):118-124. DOI: 10.1016/j.ijsrc.2018.07.001

[25] Ettema R, Constantinescu G, Melville BW. Flow-field complexity and design estimation of pier-scour depth: Sixty years since Laursen and Toch. Journal of Hydraulic Engineering. 2017;**143**(9):03117006. DOI: 10.1061/(ASCE)HY.1943-7900.0001330

[26] Sumer BM, Christiansen N, Fredsøe J. Horseshoe vortex and vortex shedding around a vertical wall-mounted cylinder exposed to waves. Journal of Fluid Mechanics. 1997;**332**: 41-70. DOI: 10.1017/S0022112096003898

[27] Williamson CHK. Sinusoidal flow relative to circular cylinders. Journal of Fluid Mechanics. 1985;**155**:141-174. DOI: 10.1017/S0022112085001756

[28] Sumer BM, Fredsøe J, Christiansen N. Scour around a vertical pile in waves. Journal of Waterway, Port, Coastal, and Ocean Engineering. 1992;**118**(1):15-31. DOI: 10.1061/(ASCE)0733-950X(1992)118:1(15)

[29] Jonsson IG. Wave boundary layers and friction factors. In: Proceedings of the Tenth Conference on Coastal Engineering - American Society of Civil Engineers; 1966. pp. 127-148. DOI: 10.1061/9780872620087.010

[30] Baykal C, Sumer BM, Fuhrman DR, Jacobsen NG, Fredsøe J. Numerical simulation of scour and backfilling processes around a circular pile in waves. Coastal Engineering. 2017;**122**: 87-107. DOI: 10.1016/j.coastaleng.2017.01.004

[31] Qi WG, Gao FP. Physical modeling of local scour development around a large-diameter monopile in combined waves and current. Coastal Engineering. 2014;**83**:72-81. DOI: 10.1016/j.coastaleng.2013.10.007

[32] Kemp PH, Simons RR. The interaction between waves and a turbulent current: Waves propagating with the current. Journal of Fluid Mechanics. 1982;**116**:227-250. DOI: 10.1017/S0022112082000445

[33] Kemp PH, Simons RR. The interaction between waves and a turbulent current: Waves propagating against the current. Journal of Fluid Mechanics. 1983;**130**:73-89. DOI: 10.1017/S0022112083000981

[34] Olabarrieta M, Medina R, Castanedo S. Effects of wave-current interaction on the current profile. Coastal Engineering. 2010;**57**:643-655. DOI: 10.1016/j.coastaleng.2010.02.003

[35] Melville BW, Sutherland AJ. Design method for local scour at bridge piers. Journal of Hydraulic Engineering. 1988; **114**(10):1210-1226. DOI: 10.1061/(ASCE)0733-9429(1988)114:10(1210)

[36] Richardson EV, Davis SR. Evaluating scour at bridges. In: Hydraulic Engineering Circular No. 18. 4th ed. U.S. Fed. Hwy. Admin., U.S. Dept. of Transp., FHWA–NHI–01–001; 2001

[37] Johnson PA. Comparison of pier-scour equations using field data. Journal of Hydraulic Engineering. 1995;**121**(8): 626-629. DOI: 10.1061/(ASCE)0733-9429(1995)121:8(626)

[38] Deng L, Cai CS. Bridge scour: Prediction, modeling, monitoring, and countermeasures. Practice Periodical on Structural Design and Construction. 2009;**15**(2):125-134. DOI: 10.1061/(ASCE)SC.1943-5576.0000041

[39] Mueller DS. Local scour at bridge piers in nonuniform sediment under dynamic conditions [PhD thesis]. Colorado: Colorado State University; 1996

[40] Sheppard DM, Melville B, Demir H. Evaluation of existing equations for local scour at bridge piers. Journal of Hydraulic Engineering. 2013;**140**(1): 14-23. DOI: 10.1061/(ASCE)HY.1943-7900.0000800

[41] Sumer BM, Hatipoglu F, Fredsøe J. Wave scour around a pile in sand, medium dense, and dense silt. Journal of Waterway, Port, Coastal, and Ocean Engineering. 2007;**133**(1):14-27. DOI: 10.1061/(ASCE)0733-950X(2007)133:1(14)

[42] Dey S, Helkjær A, Mutlu Sumer B, et al. Scour at vertical piles in sand-clay mixtures under waves. Journal of Waterway, Port, Coastal, and Ocean Engineering. 2011;**137**(6):324-331. DOI: 10.1061/(ASCE)WW.1943-5460.0000095

[43] Zanke UCE, Hsu TW, Roland A, Oscar L, Reda D. Equilibrium scour depths around piles in noncohesive sediments under currents and waves. Coastal Engineering. 2011;**58**:986-991. DOI: 10.1016/j.coastaleng.2011.05.011

[44] Sumer BM, Fredsøe J. Scour around pile in combined waves and current. Journal of Hydraulic Engineering. 2001; **127**(5):403-411. DOI: 10.1061/(ASCE)0733-9429(2001)127:5(403)

[45] Sumer BM, Fredsøe J. The Mechanics of Scour in the Marine Environment. Singapore: World Scientific; 2002. DOI: 10.1142/4942

[46] Rudolph D, Bos KJ, Luijendijk AP, et al. Scour around offshore structures—Analysis of field measurements. In: Proceedings of the 2nd International Conference on Scour and Erosion; Vol. 1; 2004. pp. 14-17

[47] Rudolph D, Bos K. Scour around a monopile under combined wave-current conditions and low KC-numbers. In: Proceedings of the Sixth International Conference on Scour and Erosion; 2006. pp. 582-588

[48] Qi WG, Gao FP. Equilibrium scour depth at offshore monopile foundation in combined waves and current. Science China Technological Sciences. 2014; **57**(5):1030-1039. DOI: 10.1007/s11431-014-5538-9

[49] Sumer BM, Petersen TU, Locatelli L, et al. Backfilling of a scour hole around a pile in waves and current. Journal of Waterway, Port, Coastal, and Ocean Engineering. 2013;**139**(1):9-23.

DOI: 10.1061/(ASCE)WW.1943-5460.0000161

[50] Heller V. Scale effects in physical hydraulic engineering models. Journal of Hydraulic Research. 2011;**49**(3): 293-306. DOI: 10.1080/00221686.2011.578914

[51] Ettema R, Kirkil G, Muste M. Similitude of large-scale turbulence in experiments on local scour at cylinders. Journal of Hydraulic Engineering. 2006; **132**(1):33-40. DOI: 10.1061/(ASCE)0733-9429(2006)132:1(33)

[52] Ettema R, Melville BW, Barkdoll B. Scale effect in pier-scour experiments. Journal of Hydraulic Engineering. 1998; **124**(6):639-642. DOI: 10.1061/(ASCE)0733-9429(1998)124:6(639)

[53] Melville BW, Chiew YM. Time scale for local scour at bridge piers. Journal of Hydraulic Engineering. 1999;**125**(1): 59-65. DOI: 10.1061/(ASCE)0733-9429(1999)125:1(59)

[54] Sheppard DM, Odeh M, Glasser T. Large scale clear-water local pier scour experiments. Journal of Hydraulic Engineering. 2004;**130**(10):957-963. DOI: 10.1061/(ASCE)0733-9429(2004)130:10(957)

[55] Lee S, Sturm T. Effect of sediment size scaling on physical modeling of bridge pier scour. Journal of Hydraulic Engineering. 2009;**135**(10):793-802. DOI: 10.1061/(ASCE)HY.1943-7900.0000091

[56] Negro V, López-Gutiérrez JS, Esteban MD, Matutano C. Uncertainties in the design of support structures and foundations for offshore wind turbines. Renewable Energy. 2014;**63**:125-132. DOI: 10.1016/j.renene.2013.08.041

[57] Qi WG, Li YX, Xu K, Gao FP. Physical modelling of local scour at twin piles under combined waves and current. Coastal Engineering. 2019;**143**: 63-75. DOI: 10.1016/j.coastaleng.2018.10.009

[58] Briaud JL, Ting FCK, Cheng HC, Gudavalli R, Perugu S, Wei GS. SRICOS: Prediction of scour rate in cohesive soils at bridge piers. Journal of Geotechnical and Geoenvironmental Engineering. 1999;**125**(4):237-246. DOI: 10.1061/(ASCE)1090-0241(1999)125:4(237)

[59] Sumer BM, Christiansen N, Fredsoe J. Time scale of scour around a vertical pile. In: Proceedings of the Second International Offshore and Polar Engineering Conference; San Francisco; 1992. pp. 308-315

[60] Soulsby R. Dynamics of Marine Sands. UK: Thomas Telford; 1997

[61] Petersen TU, Sumer BM, Fredsøe J. Time scale of scour around a pile in combined waves and current. In: International Conference on Coastal Engineering; 2012. pp. 971-988

[62] Dyrseth S. Time scales for scour below pipelines and around vertical piles in nonlinear random waves and current [Master thesis]. Trondheim: Norwegian University of Science and Technology; 2015

[63] Chen B, Li S. Experimental study of local scour around a vertical cylinder under wave-only and combined wave-current conditions in a large-scale flume. Journal of Hydraulic Engineering. 2018;**144**(9):04018058. DOI: 10.1061/(ASCE)HY.1943-7900.0001502

[64] Harris J, Whitehouse RJS, Benson T. The time evolution of scour around offshore structures. ICE-Maritime Engineering. 2010;**163**(1):3-17. DOI: 10.1680/maen.2010.163.1.3

[65] Hartvig PA, Thomsen JMC, Frigaard P, et al. Experimental study of the development of scour and backfilling. Coastal Engineering Journal.

2010;**52**(02):157-194. DOI: 10.1142/S0578563410002154

[66] Sørensen SPH, Ibsen LB, Frigaard P. Experimental evaluation of backfill in scour holes around offshore monopiles. In: Frontiers in Offshore Geotechnics II. London: Taylor & Francis Group; 2010. pp. 635-640

[67] Sørensen SPH, Ibsen LB. Assessment of foundation design for offshore monopiles unprotected against scour. Ocean Engineering. 2013;**63**: 17-25. DOI: 10.1016/j.oceaneng.2013.01.016

[68] Reese LC, Cox WR Koop FD. Analysis of laterally loaded piles in sand. In: Proceedings of the Offshore Technology Conference; Houston, TX; 1974. p. 2080

[69] O'Neill MW, Murchison JM. An Evaluation of p-y Relationships in Sands, Report to American Petroleum Institute. Houston, TX, USA: University of Houston; 1983

[70] Suryasentana SK, Lehane BM. Numerical derivation of CPT-based p-y curves for piles in sand. Géotechnique. 2014;**64**(3):186-194. DOI: 10.1680/geot.13.P.026

[71] Zhu B, Sun YX, Chen RP, Guo WD, Yang YY. Experimental and analytical models of laterally loaded rigid monopiles with hardening p-y curves. Journal of Waterway, Port, Coastal, and Ocean Engineering. 2015;**141**(6): 04015007. DOI: 10.1061/(ASCE)WW.1943-5460.0000310

[72] American Petroleum Institute. Geotechnical and Foundation Design Considerations, ANSI/API Recommended Practice 2 GEO. 1st ed; 2011

[73] Qi WG, Gao FP, Randolph MF, Lehane BM. Scour effects on p-y curves for shallowly embedded piles in sand. Géotechnique. 2016;**8**:648-660. DOI: 10.1680/jgeot.15.P.157

[74] Lin C, Bennett C, Han J, Parsons RL. Scour effects on the response of laterally loaded piles considering stress history of sand. Computers and Geotechnics. 2010;**37**:1008-1014. DOI: 10.1016/j.compgeo.2010.08.009

[75] Lin C, Han J, Bennett C, Parsons RL. Analysis of laterally loaded piles in sand considering scour hole dimensions. Journal of Geotechnical and Geoenvironmental Engineering. 2014; **140**:04014024. DOI: 10.1061/(ASCE)GT.1943-5606.0001111

[76] Yang XF, Zhang CR, Huang MS, Yuan JY. Lateral loading of a pile using strain wedge model and its application under scouring. Marine Georesources & Geotechnology. 2017;**36**(3):340-350. DOI: 10.1080/1064119X.2017.1317889

[77] Tseng WC, Kuo YS, Chen JW. An investigation into the effect of scour on the loading and deformation responses of monopile foundations. Energies. 2017;**10**(8):1190. DOI: 10.3390/en10081190

[78] Sørensen SPH. Soil-structure interaction for non-slender, large-diameter offshore monopiles [PhD thesis]. Aalborg, Denmark: Aalborg University; 2012

[79] Lin C. Wu R. Evaluation of vertical effective stress and pile lateral capacities considering scour-hole dimensions. Canadian Geotechnical Journal. 2018; **56**(1):135-143. DOI: 10.1139/cgj-2017-0644

[80] Bekken L. Lateral behavior of large diameter offshore monopile foundations for wind turbines [PhD thesis]. TU Delft: Delft University of Technology; 2009

[81] Byrne BW, McAdam R, Burd HJ, et al. New design methods for large

diameter piles under lateral loading for offshore wind applications. In: Proceedings of 3rd International Symposium on Frontiers in Offshore Geotechnics. Oslo: CRC Press; 2015. pp. 1-6

[82] Gao FP, Li JH, Qi WG, Hu C. On the instability of offshore foundations: theory and mechanism. Science China - Physics Mechanics & Astronomy. 2015; **58**(12):124701. DOI: 10.1007/s11433-015-5745-9

[83] Qi WG, Gao FP. Numerical study of local scour effects on the lateral pile-soil interaction. In: Scour and Erosion: Proceedings of the 8th International Conference on Scour and Erosion; Oxford, UK; 12–15 September 2016. pp. 293-300

[84] Poulos H, Hull T. The role of analytical geomechanics in foundation engineering. In: Foundation Engineering: Current Principles and Practices. Reston, VA: ASCE 2; 1989. pp. 1578-1606

[85] Zhang CR, Zhang X, Huang MS, Tang HW. Responses of caisson-piles foundations to long-term cyclic lateral load and scouring. Soil Dynamics and Earthquake Engineering. 2019;**119**: 62-74. DOI: 10.1016/j.soildyn.2018.12.026

[86] Achmus M, Kuo YS, Abdel-Rahman K. Numerical investigation of scour effect on lateral resistance of windfarm monopiles. In: The Twentieth International Offshore and Polar Engineering Conference. Beijing, China: International Society of Offshore and Polar Engineers; 2010. pp. 619-623

[87] Bisoi S, Haldar S. Dynamic analysis of offshore wind turbine in clay considering soil-monopile-tower interaction. Soil Dynamics and Earthquake Engineering. 2014;**63**:19-35. DOI: 10.1016/j.soildyn.2014.03.006

[88] Shadlou M, Bhattacharya S. Dynamic stiffness of monopiles supporting offshore wind turbine generators. Soil Dynamics and Earthquake Engineering. 2016;**88**:15-32. DOI: 10.1016/j.soildyn.2016.04.002

[89] Kallehave D, Byrne BW, LeBlanc Thilsted C, Mikkelsen KK. Optimization of monopiles for offshore wind turbines. Philosophical Transactions of the Royal Society A - Mathematical Physical and Engineering Sciences. 2015;**373**: 20140100. DOI: 10.1098/rsta.2014.0100

[90] Prendergast LJ, Gavin K, Doherty P. An investigation into the effect of scour on the natural frequency of an offshore wind turbine. Ocean Engineering. 2015; **101**:1-11. DOI: 10.1016/j.oceaneng.2015.04.017

[91] Prendergast LJ, Reale C, Gavin K. Probabilistic examination of the change in eigen frequencies of an offshore wind turbine under progressive scour incorporating soil spatial variability. Marine Structures. 2018;**57**:87-104. DOI: 10.1016/j.marstruc.2017.09.009

[92] Damgaard M, Andersen LV, Ibsen LB. Dynamic response sensitivity of an offshore wind turbine for varying subsoil conditions. Ocean Engineering. 2015;**101**:227-234. DOI: 10.1016/j.oceaneng.2015.04.039

[93] Damgaard M, Ibsen LB, Andersen LV, Andersen JK. Cross-wind modal properties of offshore wind turbines identified by full scale testing. Journal of Wind Engineering and Industrial Aerodynamics. 2013;**116**: 94-108. DOI: 10.1016/j.jweia.2013.03.003

[94] Schafhirt S, Page A, Eiksund GR, Muskulus M. Influence of soil parameters on the fatigue lifetime of offshore wind turbines with monopile support structure. Energy Procedia. 2016;**94**:347-356. DOI: 10.1016/j.egypro.2016.09.194

[95] Rezaei R, Duffour P, Fromme P. Scour influence on the fatigue life of operational monopile-supported offshore wind turbines. Wind Energy. 2018;**21**(9):683-696. DOI: 10.1002/we.2187

[96] Lehane BM, Pedram B, Doherty JA, Powrie W. Improved performance of monopiles when combined with footings for tower foundations in sand. Journal of Geotechnical and Geoenvironmental Engineering. 2014; **140**:04014027. DOI: 10.1061/(ASCE)GT.1943-5606.0001109

[97] Anastasopoulos I, Theofilou M. Hybrid foundation for offshore wind turbines: Environmental and seismic loading. Soil Dynamics and Earthquake Engineering. 2016;**80**: 192-209. DOI: 10.1016/j.soildyn.2015.10.015

[98] Stone KJL, Arshi HS, Zdravkovic L. Use of a bearing plate to enhance the lateral capacity of monopiles in sand. Journal of Geotechnical and Geoenvironmental Engineering. 2018; **144**(8):04018051. DOI: 10.1061/(ASCE)GT.1943-5606.0001913

[99] Wang X, Zeng X, Li X, Li J. Investigation on offshore wind turbine with an innovative hybrid monopile foundation: an experimental based study. Renewable Energy. 2019;**132**: 129-141. DOI: 10.1016/j.renene.2018.07.127

[100] Whitehouse RJS, Sutherland J, Harris JM. Evaluating scour at marine gravity foundations. In: Proceedings of the ICE-Maritime Engineering; 2011. pp. 143-157

[101] Qi WG, Gao FP. Wave induced instantaneously-liquefied soil depth in a non-cohesive seabed. Ocean Engineering. 2018;**153**:412-423. DOI: 10.1016/j.oceaneng.2018.01.107

[102] Qi WG, Li CF, Jeng DS, Gao FP, Liang Z. Combined wave-current induced excess pore-pressure in a sandy seabed: Flume observations and comparisons with theoretical models. Coastal Engineering. 2019;**147**: 89-98. DOI: 10.1016/j.coastaleng.2019.02.006

[103] Lin Z, Pokrajac D, Guo Y, Jeng DS, Tang T, Rey N, et al. Investigation of nonlinear wave-induced seabed response around mono-pile foundation. Coastal Engineering. 2017;**121**:197-211. DOI: 10.1016/j.coastaleng.2017.01.002

[104] Zhao HY, Jeng DS, Liao CC, Zhu JF. Three-dimensional modeling of wave-induced residual seabed response around a mono-pile foundation. Coastal Engineering. 2017;**128**:1-21. DOI: 10.1016/j.coastaleng.2017.07.002

[105] Zhu B, Ren J, Ye GL. Wave-induced liquefaction of the seabed around a single pile considering pile-soil interaction. Marine Georesources & Geotechnology. 2018;**36**(1):150-162. DOI: 10.1080/1064119X.2017.1340374

[106] Duan LL, Jeng DS, Wang D. PORO-FSSI-FOAM: Seabed response around a mono-pile under natural loadings. Ocean Engineering. 2019;**184**: 239-254. DOI: 10.1016/j.oceaneng.2019.05.024

[107] Sui TT, Zhang C, Guo YK, Zheng JH, Jeng DS, Zhang JS, et al. Three-dimensional numerical model for wave-induced seabed response around mono-pile. Ships and Offshore Structures. 2016;**11**(6):667-678. DOI: 10.1080/17445302.2015.1051312

[108] Li Y, Ong MC, Tang T. Numerical analysis of wave-induced poro-elastic seabed response around a hexagonal gravity-based offshore foundation. Coastal Engineering. 2018;**136**:81-95. DOI: 10.1016/j.coastaleng.2018.02.005

Chapter 5

Frost Heave Deformation Analysis Model for Microheave Filler

Ye Yangsheng, Du Xiaoyan, Zhang Qianli and Chai Jinfei

Abstract

With the rapid development of high-speed railway, high-speed railways pose new requirements on subgrade frost heave deformation control. Microheave in conventional non-frost-heave filler cannot meet the requirements of high-speed railways for high levels of smoothness and stability and threaten high-speed train operation safety. To solve problems of seasonal permafrost region subgrade filler microheave in China, combined laboratory test and theoretical analysis, this research analyzed the physical properties of frost heave influencing factors for microheave filler. The influence of skeleton grain during frost heave formation is revealed. The microheave filler frost heave development mechanism is investigated. On this basis, based on the principle of minimum energy, a frost heave calculation formula for microheave filler is deduced, and a frost heave deformation analysis model for microheave filler is created. In addition, the effectiveness of the model is demonstrated in an indoor test. This study provides a theoretical reference for controlling the frost heaving deformation of railway subgrade.

Keywords: high-speed railway subgrade, microheave filler, frost heave influencing factor, frost heave development mechanism, frost heave analysis model

1. Introduction

China now has the world's longest high-speed railway operation mileage, fastest operational speed and largest scale of projects under construction. By the end of 2016, the high-speed railway operational mileage had reached 2.2×10^4 km, which was the largest in the world. Compared with normal speed railways, the major features of high-speed railways are that they are fast, comfortable and safe. For ballasted track, high-speed railway subgrade settlement should not exceed 5 or 3 cm in transition segments. For ballastless track high-speed railway subgrades, the ideal goal in actual projects is zero settlement, in particular for deformation-sensitive sections such as transit segments, for which settlement is controlled to be under 5 mm and to not exceed 1/1000 at corners. To achieve the above goal, track structure smoothness and stability should be guaranteed, which is reliant on providing track structures with subgrade structures of high strength, high rigidity, uniform longitudinal variation, high stability and durability. Currently, in northeast China, the Harbin-Dalian, Panjin-Yingkou, Harbin-Qiqihar, Shenyang-Dandong and Jilin-Hunchun high-speed railways span seasonal permafrost regions. Compared with high-speed railways in non-frost regions, for high-speed railways in permafrost regions, subgrade filler frost heave induces uneven settlement, which has become a critical issue [1]. In northern China, all railway subgrades have developed various levels of frost heave, which directly threatens travelling safety [2].

In China, both existing high-speed trains and those under construction follow the guidelines in the *Code for Design of Special Subgrade of Railway* (TB10035-2006) regarding filler selection and the design of scattered row impermeable and anti-frost-heave structures. That is, the standard for filler frost heave control is 1%. However, on high-speed railway subgrades in Northeast China, the maximum depth of frost reaches 3 m. According to this standard, subgrade frost heave deformation cannot meet the requirement for ballastless track deformation control. In addition, in the *Code for Design of Special Railway Subgrades*, the seasonal frozen soil frost heave grading object is natural frozen soil, not manually compacted subgrade filler. Therefore, the microheave deformation mechanism for subgrade filler should be investigated, and a microheave deformation analysis model should be created to provide a theoretical basis for specification development and engineering design.

Currently, subgrade seasonal frost heave problems worldwide are investigated via experiments and theoretical models. In an experimentally based investigation, Askar and Zhanbolat investigated the effects of freeze-thaw cycles on a high-speed highway subgrade; a frost heave test was performed using field soil samples to identify frost heave and frost heave pressure in regional soil samples and frost heave depth under different conditions. They suggested that frost heave could be effectively reduced by replacing frost heave sensitive soil with coarse grained soil and deploying a drainage system [3].

Konrad suggested that it was inadequate to determine if subgrade filler had an anti-frost property by grain grading alone; the frost heave properties of fine grains should also be considered [4]. In addition, the frost heave sensitivities of subgrade fillers with superior grading increase linearly with fine grain content and active mineral content [5]. Bilodeau et al. investigated the effects of grain grading and fine grain properties on frost heave sensitivity and found that when fine grain content was low, fine grain mineral composition and grading had greater impacts on frost heave sensitivity. In addition, fine grains with different uniformity coefficients resulted in different frost heave levels [6]. Uthus et al. suggested that the main factors that influenced frost heave were water and fine grains (active minerals). Among these factors, water had greater impacts on frost heave than the material elastic modulus [7].

Wang et al. suggested that frost heave sensitivity and strength of coarse grained soil were affected by fine grain content. Frost heave tests and triaxial tests showed that when the fine grain soil content was 5%, coarse grained soil had weak frost heave property but high shear strength [8]. Bilodeau et al. investigated the relation between freeze-thaw cycle and long-term compression-induced permanent displacement in subgrade coarse grained materials. The results showed that pressure had a greater influence in a saturated soil sample that had not undergone freeze-thaw cycling. In addition, freeze-thaw cycles had a significant influence [9]. Guo et al. performed a frost heave test in a closed system with various levels of water content and compactness as well as in an open system, investigated the Lanzhou-Xinjiang railway subgrade soil frost heave characteristics and reached a conclusion that due to water content replenishment, the frost heave rate in the open system significantly exceeded that in the closed system [10].

In a model-based investigation, Sheng et al. undertook a model test and discovered that train cycle load increased subgrade soil pore hydro-pressure; under the pump suction effect, water level was observed to rise to the frost heave layer, which resulted in continuous frost heave in a high-speed railway subgrade [11]. Abdalla et al. making reasonable assumptions, extended a porosity function model from a frost heave forecast to a freeze-thaw cycle simulation [12]. Zhang and Michalowski introduced a THM (thermal-hydro-mechanical) model for frost heave sensitive soils such as viscous soils to describe temperature, hydraulic pressure and force field changes induced by freeze-thaw cycling and subsequent physical phenomena such as expansion and settlement; they also provided a numerical simulation for the one-dimensional problem [13]. Bronfenbrener described heat and matter transfer in a fine grain porous medium with frost heave during a phase transition and provided a model for the problem with a movement boundary condition. Comparison with test data proved that within a certain level of precision, the model was viable for frost heave estimation [14]. Yin created a seasonal permafrost region, subgrade soil, multi-factor, frost heave rate forecast model via field measurements and theoretical analysis and applied it to a project [15].

In this paper, based on a field survey in a permafrost region and high-speed railway subgrade filler and frost heave characteristics, the concept of subgrade microheave filler is proposed. A frost heave calculation formula for a microheave filler in closed systems is deduced via an indoor test and theoretical analysis. A frost heave deformation analysis model for microheave filler is created, and its effectiveness is proven.

2. Experimental study on microheave filler frost heave influencing factors

Upon conducting an experimental study, Wang et al. suggested that there were three categories of railway subgrade filler: over coarse grained soils, coarse grained soils and fine grain soils [16]. In this paper, the large amounts of coarse grained soil filler in high-speed railway subgrades in the seasonal permafrost region of China are called microheave filler.

The bed filler used in high-speed railway subgrades in the permafrost regions of China is arranged as follows. A normal embankment bed surface is filled with a layer of 55-cm grading gravel, below which are 5 cm of thick coarse sand and 2.1 m of thick soil in two groups, A and B. In the frozen depth range, the embankment bed is filled with non-frost-heave soil in groups A and B, and soils in groups A, B and C exist below the beds. The top 1.0 m of the embankment bed floor is filled with non-frost-heave filler in groups A and B. Based on the distribution of filler along the line, the lower sections have filler in groups A and B. Low embankment bed surfaces are filled with grading gravel. Non-frost-heave filler in groups A and B is used in the range of the upper meter of bed floors. The lower sections are filled with filler in groups A and B. In the bed floor lower sections, the grading gravel layer water content is 3.2–5.9%, and the fine grain content is 3.2–10.7%. In lower sections, the non-frost-heave soil filler water content generally is 3.7–9.2%, and the fine grain content is 0.4–9.3%. In seasonal permafrost regions, the entire beds are excavated and replaced. The bed surfaces are filled with grading gravel, the upper 1 m of the bed floors are filled with non-frost-heave filler in groups A and B, and the lower sections are filled with filler in groups A and B. The grading gravel layer water contents are 2.2–3.2%, the fine grain contents are 8.0–15.1%, the water contents in the lower section non-frost-heave soil filler are generally 3.7–9.2%, and the fine grain contents are 10.9–16.9%.

Coarse grained soil frost heave tests and analysis proved that the factors that influence major coarse grained soil frost heave include the soil property, water content, temperature and load [17]. To investigate the characteristics of high-speed railway subgrade filler frost heave in permafrost regions, in this chapter, typical permafrost region subgrade soil samples were selected to perform soil property tests for filler covering water content, liquid plastic limit and grain analysis. On this basis, the effects of various factors on filler frost heave were analyzed to obtain rules regarding the influence of soil fine grain content, compactness, water content, permeability and coarse grain pore filling by fine grains on the properties of coarse grained soil frost heave. It is worth noting that the test reported in this section was based on the conventional liquid and plastic limit combined determination method. In the test, fine grain soil liquid limit water and plastic limit water contents were measured to calculate the soil plasticity index and liquid index, which were used as a basis to evaluate the properties of filler soils. A standard screening test was employed to calculate the relative content of each grain group and to determine soil grain composition. In addition, for subgrades, compactness is not only a control index that ensures subgrade filling quality but also a major factor that affects the properties of filler frost heave. Under identical conditions, as compactness increases, fine grain soil frost heave increases initially and then decreases.

3. Microheave filler frost heave test

3.1 Test apparatus

The frost heave test apparatus consisted of a specimen box, an incubator, a temperature control system, a temperature monitoring system, a deformation monitoring system, a pressure loading system and a water supply system, as shown in **Figure 1**. The specimen box had a diameter of 15 cm and a height of 15 cm.

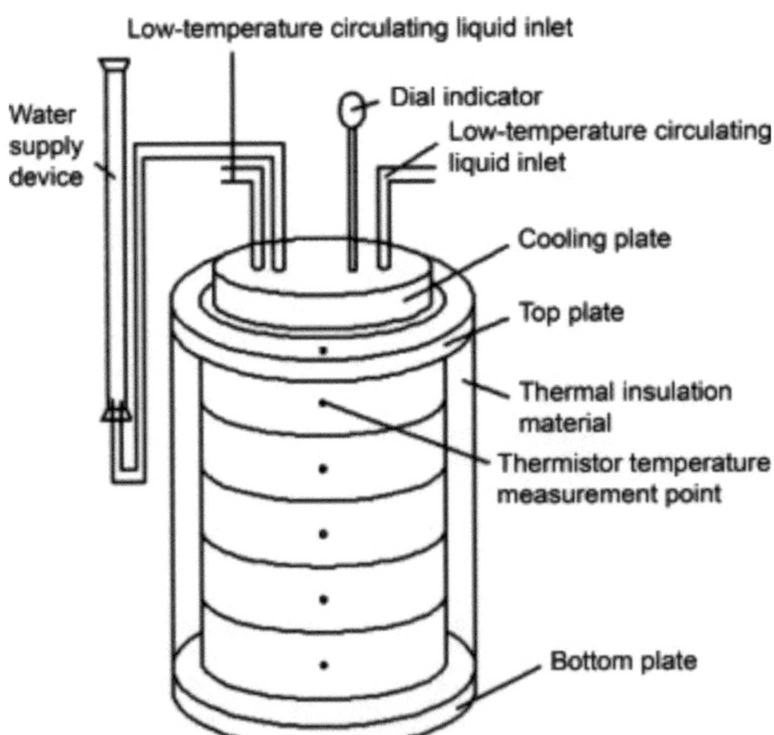

Figure 1.
Frost heave test apparatus.

3.2 Test method

Proper amounts of dry soil samples were mixed with water to produce the required water contents. The soil samples were separated into layers of predefined compactness (5 layers, each with thicknesses of 3 cm) and placed in the specimen box. The specimen box with a sample was placed in the incubator. Thermistor thermometers were deployed on the specimen top plate, bottom plate and side. The specimen box was wrapped in 5 cm of thick plastic foam for heat insulation. The top plate and bottom plate temperatures were controlled using two high precision low temperature circulation cooling systems. A dial indicator was installed on the top plate to monitor specimen deformation. Finally, the temperatures for the specimens were collected automatically via the data collector. The testing was performed in a closed environment. The unidirectional freezing method was applied [18], the duration of the freezing process for each specimen was 72 h, and the direction of freezing was from top to bottom. At the beginning of the tests, the soil column temperature was stabilized at approximately 1°C and maintained for 6 h. The bottom plate temperature was then maintained at 1°C. At 0.5 h, the top plate temperature was decreased to −15°C, and the soil samples were frozen rapidly from the top surface. The top plate temperature was increased to −2°C, and top plate temperature was then decreased by a certain amount per hour. At the conclusion of each test, the soil samples were separated into layers in the low temperature incubator, and the water contents were measured.

3.3 Test items

Test filler was obtained from the subgrade filling soil of a high-speed railway in Northeast China. The grades are listed in **Table 1**. The filler plastic limit ω_p was 19.2%, and the optimal water content ω_0 was 13.2%.

To investigate the effects of factors such as water content, filler and external load on microheave filler frost heave for this paper, a frost heave experimental study was conducted to investigate the following aspects.

1. Effects of filler plastic limit and optimal water content on microheave filler frost heave.

Proper amounts of dry filler were mixed with various amounts of water to produce 30 groups of ω specimens with various levels of initial water content. The initial water content values of the 30 groups of specimens are listed in **Table 2**. The

Grain size range (mm)	content (%)
31.5–45.0	20
22.4–31.5	10
7.1–22.4	25
1.7–7.1	20
0.5–1.7	10
0.1–0.5	9
0.075–0.1	3
<0.075	3

Table 1.
Test filler grades.

Specimen	Initial water content ω(%)
1	12.3
2	14.5
3	15.1
4	15.5
5	15.9
6	16.0
7	16.5
8	18.0
9	18.5
10	18.8
11	19.0
12	19.6
13	20.0
14	20.5
15	21.3
16	13.3
17	14.3
18	14.4
19	14.7
20	15.5
21	15.6
22	15.8
23	16.2
24	16.3
25	17.5
26	17.8
27	18.0
28	19.1
29	19.4
30	21.7

Table 2.
Specimen initial water contents.

first 15 groups of specimens were used to investigate the effects of filler plastic limit on the microheave filler frost heave property, and the next 15 groups of specimens were used to investigate the effects of filler optimal water content on the microheave filler frost heave property. The required amounts of the specimens were calculated based on a compactness of 0.95. The specimens were then compacted in the specimen box for frost heave testing.

2. Effects of volumetric water content on microheave filler frost heave

Proper amounts of dry filler were mixed with various amounts of water to produce 15 groups of ω_V specimens with various levels of initial water content. The initial water content values of the 15 groups of specimens are listed in **Table 3**. The required amounts of the specimens were calculated based on a compactness of 0.95. The specimens were then compacted in the specimen box for frost heave testing.

3. Effects of filler content and filling rate on microheave filler frost heave

To prepare specimens with the required filler contents, proper amounts of dry filler were mixed with the required filler. In addition, the specimen's filler volume fill rate s was calculated via the following formula.

$$s = \frac{1\text{-}V_1}{V_2} \times 100\% \tag{1}$$

where $V_1 = \frac{M_1}{\gamma_1}$, $V_2 = \frac{M_2}{\gamma_2}$.

V_1 is the filler skeleton grain volume fraction, V_2 is the filler volume fraction, M_1 is the skeleton grain mass for a unit specimen volume, γ_1 is the skeleton grain dry density for the corresponding soil sample compactness, M_2 is the filler mass for a unit specimen volume, and γ_2 is the filler dry density for the corresponding soil sample compactness.

In this test, 15 groups of specimens were prepared. The specimen filler contents and filler filling rates are listed in **Table 4**. In the test, fillers with different contents were used to ensure that the filler skeleton component was proportionally adjusted under the premise that the total mass of filler remained unchanged. All the specimens used in the frost heave test had 15% water content and compactness values of 0.95.

4. Effects of filler frost heave on microheave filler frost heave.

Filler frost heave and pure filler frost heave tests were performed for the 15 groups of filler content specimens prepared in (3) under test conditions of 15% water content and 0.95 compactness [19].

specimen	volumetric water content ωV(%)
1	5.83
2	7.02
3	8.30
4	9.60
5	11.07
6	11.34
7	11.85
8	13.17
9	14.48
10	15.27
11	15.80
12	16.04
13	16.57
14	17.49
15	18.21

Table 3.
Specimen volumetric water contents.

Specimen	Filler content (%)	Filler filling rate (%)
1	1.0	0.056
2	2.0	0.118
3	2.5	0.198
4	2.8	0.272
5	3.0	0.316
6	3.2	0.334
7	5.0	0.363
8	5.5	0.381
9	6.0	0.445
10	9.0	0.465
11	11.0	0.529
12	16.0	0.532
13	18.0	0.606
14	40.0	0.859
15	45.0	0.928

Table 4.
Specimen filler content and filler filling rate.

Figure 2.
Loading apparatus used for the frost heave test.

5. Effects of overlying load on microheave filler frost heave.

A specimen with a 10% filler content was prepared for the frost heave test under test conditions in which the initial water content was 15% and the overlying loads were 5, 10, 20, 30, 40, 55, 65 and 80 kPa. **Figure 2** shows the loading apparatus used in the frost heave test.

4. Analysis of test results

The microheave filler frost heave rate η was calculated via the following formula:

$$\eta = \frac{\Delta h}{H_f} \times 100\% \quad (2)$$

where Δh is the overall specimen frost heave in mm, and H_f is the frozen depth (excluding frost heave) in mm.

4.1 Effects of water content on microheave filler frost heave

1. Relation between filler plastic limit, optimal water content and microheave filler frost heave.

Figure 3 shows the relation between ω-ω_p (difference between the filler initial water content ω and filler plastic limit water content ω_p) and filler frost heave rate η. **Figure 4** shows the relation between ω-ω_0 (difference between the filler initial water content ω and filler optimal water content ω_0) and filler frost heave rate η.

Figures 3 and **4** show that for filler water contents of $\omega \leq \omega_p + 2$ or $\omega \leq \omega_o + 4.6$, the frost heave rate $\eta < 1\%$. However, as the water content increased further, the filler frost heave rate increased significantly.

1. Relation between microheave filler volumetric water content and frost heave rate.

Figure 5 shows the relation between the filler volumetric water content ω_v and filler frost heave rate η.

Figure 5 shows that when $\omega_v \leq 13\%$, the filler frost heave rate was insensitive to increases in the volumetric water content, and when $\omega_v > 13\%$, filler frost heave increased significantly with increasing volumetric water content.

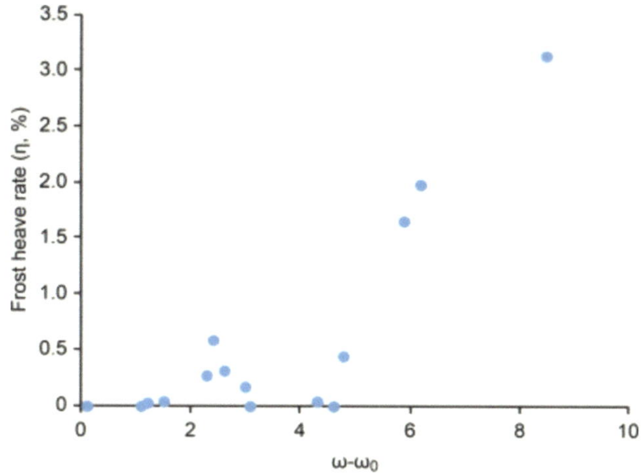

Figure 3.
Relation between the filler plastic limit and filler frost heave.

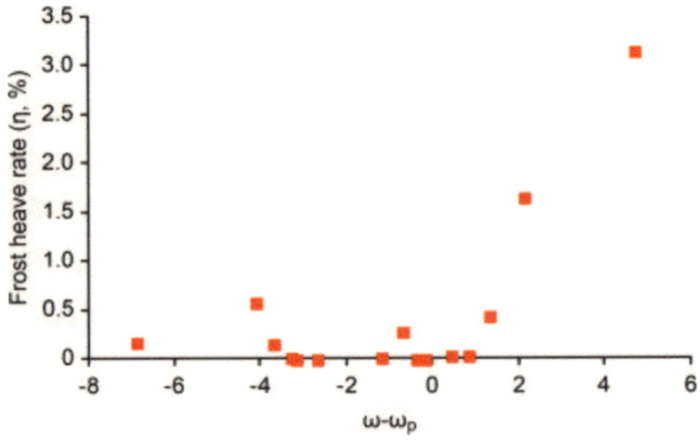

Figure 4.
Relation between the filler optimal water content and filler frost heave.

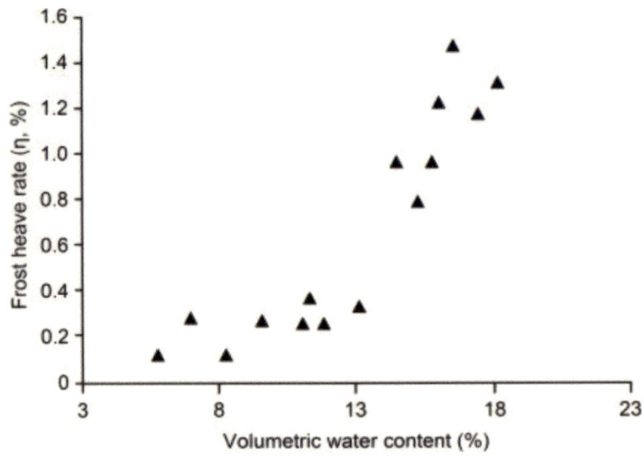

Figure 5.
Relation between the filler volumetric water content and filler frost heave rate.

4.2 Effects of filler on microheave filler frost heave

1. Relation between filler content, filling rate and microheave filler frost heave.

Figure 6 shows the microheave filler frost heave rates for different filler contents.

Figure 6 shows that filler frost heave increased gradually with filler content. When the filler content was under 3%, the frost heave rate was approximately 0.2%, when the filler content was under 15%, the frost heave rate was under 1.0%, and when the filler content exceeded 15%, the filler frost heave sensitivity increased significantly with filler content.

Figure 7 shows the microheave filler frost heave rates for different filler filling rates.

Figure 7 shows that for filler filling rates below 0.18, the filler frost heave rates were under 0.2%, and when the filler filling rates were under 0.25, the filler frost heave rates were under 0.5%; filler frost heave was insensitive to increases in filler content. For filling rates exceeding 0.25, the filler frost heave rates increased significantly with the filling rate, and frost heave became sensitive. For the filling rates were under 0.37, the frost heave rates were under 1.0%.

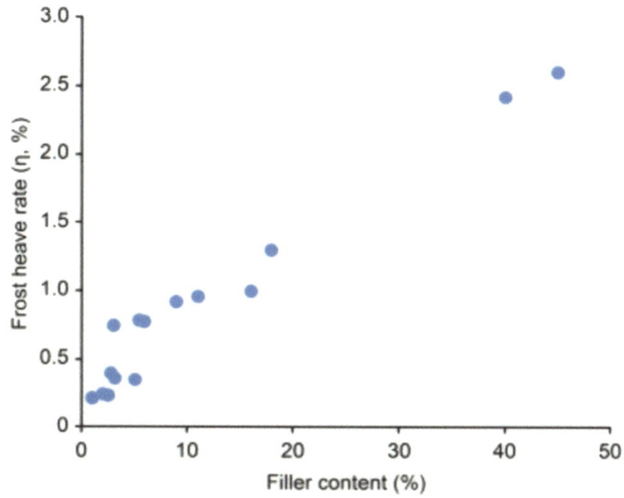

Figure 6.
Microheave filler frost heave rate for different filler contents.

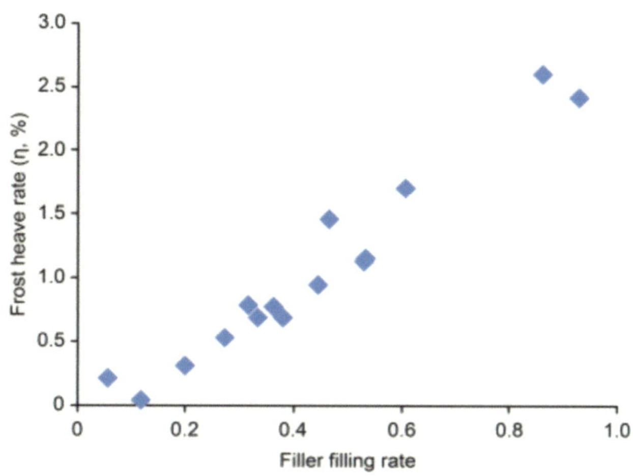

Figure 7.
Microheave filler frost heave rate for different filler filling rate.

2. Relation between filler frost heave and microheave filler frost heave.

Based on the test results, the relation between specimen filler frost heave and microheave filler frost heave for each group is shown in **Figure 8**.

Figure 8 shows that when filler frost heave was below 25 cm^3, the microheave filler exhibited no frost heave. Under this condition, filler frost heave filled pores, and there was no macroscopic filler frost heave. When filler frost heave exceeded 25 cm^3, macroscopic frost heave in the microheave filler began and increased significantly with filler frost heave.

4.3 Relation between overlying load and microheave filler frost heave

Overlying loads affect microheave filler frost heave properties in two ways: the freezing point drops with increasing external load, and overlying loads result in

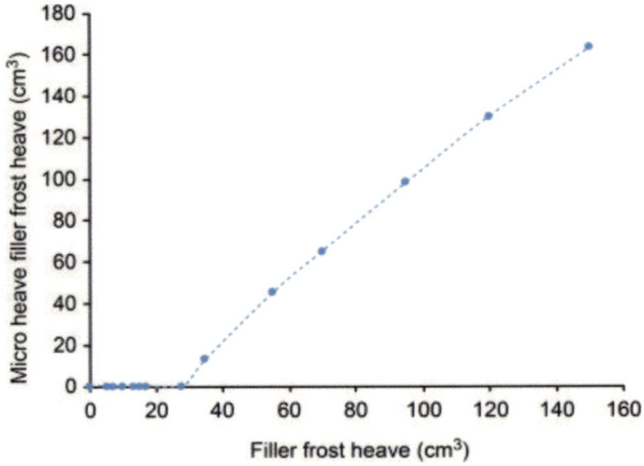

Figure 8.
Filler frost heave versus microheave filler frost heave.

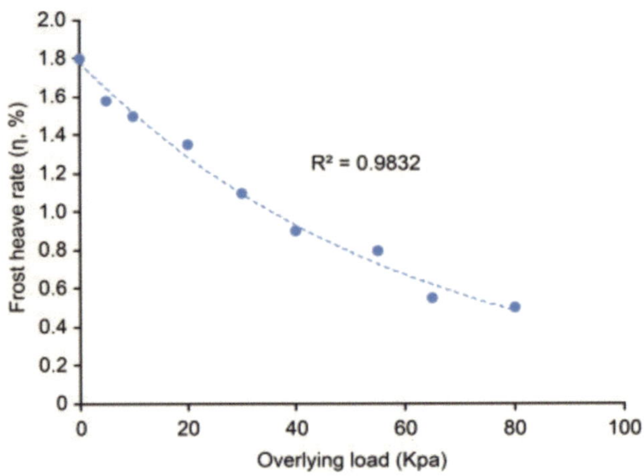

Figure 9.
Relation between frost heave rate and overlying load.

redistributions of filler water content [20]. **Figure 9** shows the microheave filler frost heave rates under different overlying loads.

Figure 9 shows that overlying load and the microheave filler frost heave rate were exponential related. The microheave filler frost heave rate decreased gradually with increasing overlying load.

The indoor test described above provided data for a qualitative analysis of microheave filler frost heave rules. To further investigate frost heave deformation, theoretical analysis and microheave filler frost heave model creation are required.

5. Microheave filler frost heave model

5.1 Microheave filler complete filling analysis

When a microheave filler specimen develops frost heave, the coarse grains do not develop frost heave and act as rigid bodies. By comparison, fine grains develop frost heave at low temperatures. Fine grain component volumes consist of soil grain and water content volumes. When fine grains develop frost heave, some water content escapes.

In this section, a simulation embodying the direct expansion of filler grains under natural conditions is examined to study filler grain expansion due to water adsorption. To investigate the complete filling of microheave filler, the mixed material specimen model was simplified. The model was assumed to consist of coarse skeletons, filler grains and pores, as shown in **Figures 10** and **11**.

Filler and residual pores constituted the mixed material skeleton pores. Among them, filler fills the skeleton pores, expands at low temperature, and, under the side constraint condition, fills the residual pores first. When the residual pores are filled, the expansion is represented as a mixed material specimen macroscopic expansion.

$$\Delta V_{mixed\ material} = \Delta V_{filler} - \Delta V_{skeleton\ pore} \tag{3}$$

If α represents the filler expansion rate, θ represents the filler content, ρ_{mixed} represents the mixed material compactness, and $G_{skeleton}$ represents the skeleton grain ratio, then

$$\Delta V_{skeleton\ pore} = \Delta V_{residual\ pore} \tag{4}$$

From formula (4) and $\Delta V_{filler} = V_{filler}\alpha$,

$$\Delta V_{mixed\ material} = \Delta V_{filler} - \Delta V_{residual\ pore} = \frac{m_{filler}}{\rho_{filler}}(1+\alpha) + \left(\frac{m_{skeleton}}{\rho_{skeleton}}\right) - V_{mixed\ material} \tag{5}$$

Because

$$m_{filler} = \frac{\theta}{(1-\theta)} m_{skeleton} \tag{6}$$

based on the macroscopic volume expansion rate,

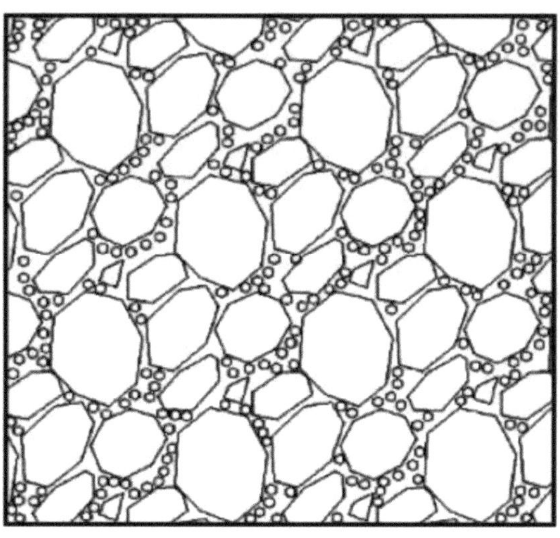

Figure 10.
Mixed material.

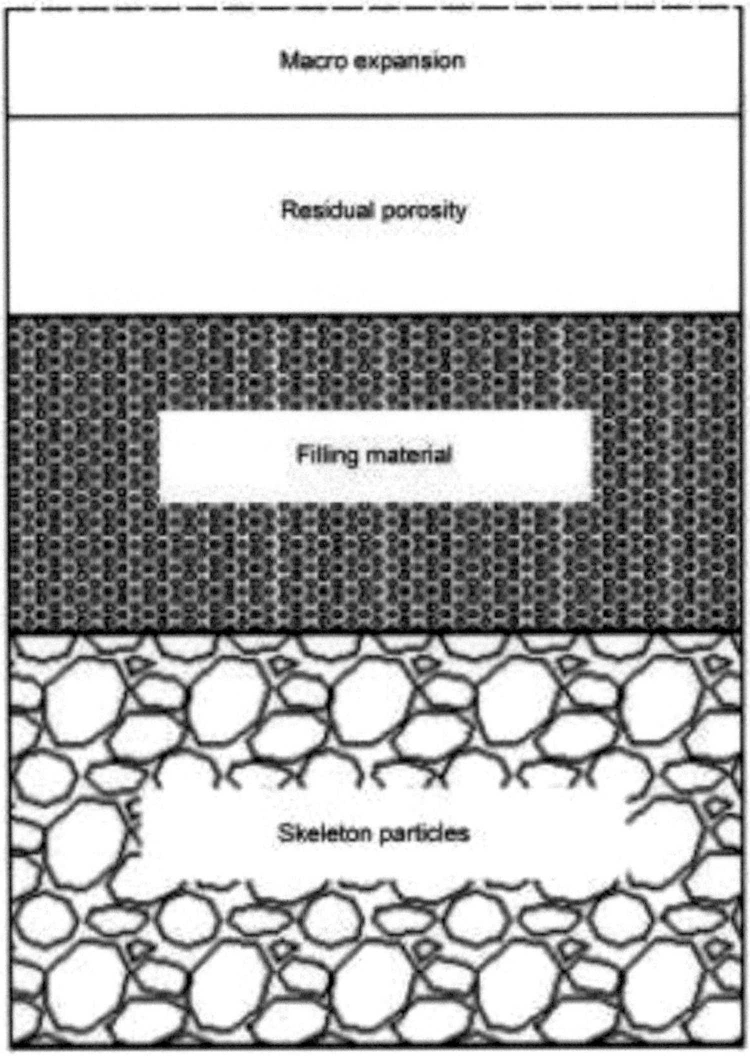

Figure 11.
Composition of the mixed material.

$$\eta_{mixed\ material} = \frac{\frac{\Delta V_{mixed\ material}}{\pi R^2}}{\frac{V_{mixed\ material}}{\pi R^2}} = \frac{\Delta V_{mixed\ material}}{V_{mixed\ material}} \quad (7)$$

$$\eta_{filler} = \alpha \quad (8)$$

From formulas (7) and (8),

$$\eta_{mixed\ material} = m_{mixed\ material}\left[\frac{\frac{\theta}{(1-\theta)}(1+\alpha)}{\rho_{filler}} + \frac{1}{\rho_{skeleton}}\right]/V_{mixed\ material} - 1 \quad (9)$$

Because

$$\rho_{mixed} = \frac{m_{skeleton} + m_{mixed\ material}}{V_{mixed\ material}} \quad (10)$$

Based on formulas (6) and (10),

$$m_{skeleton} = \rho(1-\theta)V_{mixed\ material} \tag{11}$$

Combining formulas (9) and (11),

$$\begin{aligned}\eta_{mixed\ material} &= m_{mixed\ material}\left[\frac{\frac{\theta}{(1-\theta)}(1+\alpha)}{\rho_{filler}} + \frac{1}{\rho_{skeleton}}\right]/V_{mixed\ material} - 1 \\ &= \rho_{mixed}(1-\theta)\left[\frac{\frac{\theta}{(1-\theta)}(1+\alpha)}{\rho_{filler}} + \frac{1}{\rho_{skeleton}}\right] - 1\end{aligned} \tag{12}$$

Because the skeleton grain ratio is $G_{skeleton} = \rho_{skeleton}/\rho_w$, from formula (12),

$$\eta_{mixed\ material} = \rho_{mixed}(1-\theta)\left[\frac{\frac{\theta}{(1-\theta)}(1+\alpha)}{\rho_{filler}} + \frac{1}{G_{skeleton}\rho_w}\right] - 1 \tag{13}$$

Based on the definition of a mixed material skeleton pore, $e = V_{residual\ pore}/V_S$,

$$\rho_{mixed} = \frac{m_{skeleton} + m_{mixed\ material}}{V_{skeleton} + V_{mixed\ material} + V_{residual\ pore}} = \frac{m_{skeleton} + m_{mixed\ material}}{(1+e)(V_{skeleton} + V_{mixed\ material})} \tag{14}$$

and employing the skeleton grain ratio,

$$\rho_{filler} = \frac{\theta}{\frac{1}{(1+e)\rho_{mixed}} - \frac{(1-\theta)}{(G_{skeleton}\rho_w)}} \tag{15}$$

Based on the definition of the porosity ratio,

$$e = \frac{d_s\rho_w(1+0.001w)}{\rho} - 1 = \frac{d_s\rho_w(1+0.001w)}{\rho_{mixed}} - 1 \tag{16}$$

Based on formulas (15) and (16),

$$\rho_{filler} = \frac{\theta}{\frac{1}{(1+e)\rho_{mixed}} - \frac{(1-\theta)}{(G_{skeleton}\rho_w)}} = \frac{\theta}{\frac{1}{d_s\rho_w(1+0.001w)} - (1-\theta)/(G_{skeleton}\rho_w)} \tag{17}$$

Based on formulas (15) and (17),

$$\begin{aligned}\eta_{mixed\ material} &= \rho_{mixed}(1-\theta)\left[\frac{\frac{\theta}{(1-\theta)}(1+\alpha)}{\rho_{filler}} + \frac{1}{G_{skeleton}\rho_w}\right] - 1 \\ &= \left[\frac{\rho_{mixed}(1+\alpha)}{d_s\rho_w(1+0.001w)} - \frac{\rho_{mixed}(1-\theta)\alpha}{G_{skeleton}\rho_w}\right] - 1\end{aligned} \tag{18}$$

Formula (18) is simplified to

$$\eta_{mixed\ material} = \eta(\rho_{mixed}, \theta, \alpha, d_s, \rho_w, w, G_{skeleton}) \tag{19}$$

Based on formula (19) and an analysis of the situation wherein the microheave filler is completely filled, it was found that mixed material macroscopic expansion is closely related to mixed material compactness $\eta_{mixed\ material}$, filler content θ, filler expansion rate α, mixed material ratio d_s, water compactness ρ_w, microheave filler water content w and skeleton ratio $G_{skeleton}$. Actual frost heave rates are greater than the calculated values.

5.2 Microheave filler partial filling

To further investigate the mixed material frost heave rule, microscopic analysis is required to understand the interactions between pore filler and microheave filler. During filler expansion, mixed material skeleton pores cannot be filled completely. Microheave filler specimens would expand under the side constraint condition. During expansion, the filler would experience plastic deformation and fill skeleton pores, which would result in microheave filler specimen macroscopic elevation. The manifestation of core filler stress is very complex. First, filler grains are not uniformly distributed in all skeleton pores of mixed materials. When stresses on fine aggregates are small, coarse aggregates are not affected, and coarse aggregate skeletons experience no deformation; coarse aggregate and fine aggregate contact surfaces are equivalent to the displacement boundary condition. As the filler expands and fills the remaining part of the skeleton pores, although the fine aggregate stresses are significant, because the expanded filler does not fill all the skeleton pores in the mixed material, the mixed material skeleton grains are not lifted, and the mixed material skeleton does not swell. To summarize, fine aggregates are not completely surrounded by coarse aggregates, pores are interconnected, fine aggregates either expand and deform in large pores or are squeezed to adjacent pores. Therefore, as shown in **Figure 12**, stress state cannot be described via simple mechanical models.

Based on Griffith micro-fracture theory and the minimum energy principle, a calculation formula was deduced via mathematical and mechanical analysis to undertake a quantitative analysis of the effects of rock and soil rheological properties for engineering. Filler was pushed into a cylinder using a porous piston. An analogy from physics was applied, and the filler expansion stress status was compared to the mathematical physics model in **Figure 13**.

Each pore in the microheave filler skeleton was treated as a container, as shown in **Figure 13**. The container top plate was porous, and more holes implied greater skeleton porosity.

The filling rate reflects the filling degree, which is defined as the ratio of filler volume to skeleton pore volume. Assuming that the mixed material skeleton grain pore filling rate by filler grains is β,

$$\beta = \frac{V_{filler}}{V_{skeleton\ pore}} \tag{20}$$

For an initial filling rate β_0, employing formula (20) yields

$$V_{skeleton\ pore} = \frac{V_{filler}}{\beta} \tag{21}$$

When filler fills skeleton pores, the mixed material skeleton grain volume does not change during expansion. Mixed material macroscopic volume changes can only be achieved by changing the skeleton pores, as shown by formula (22).

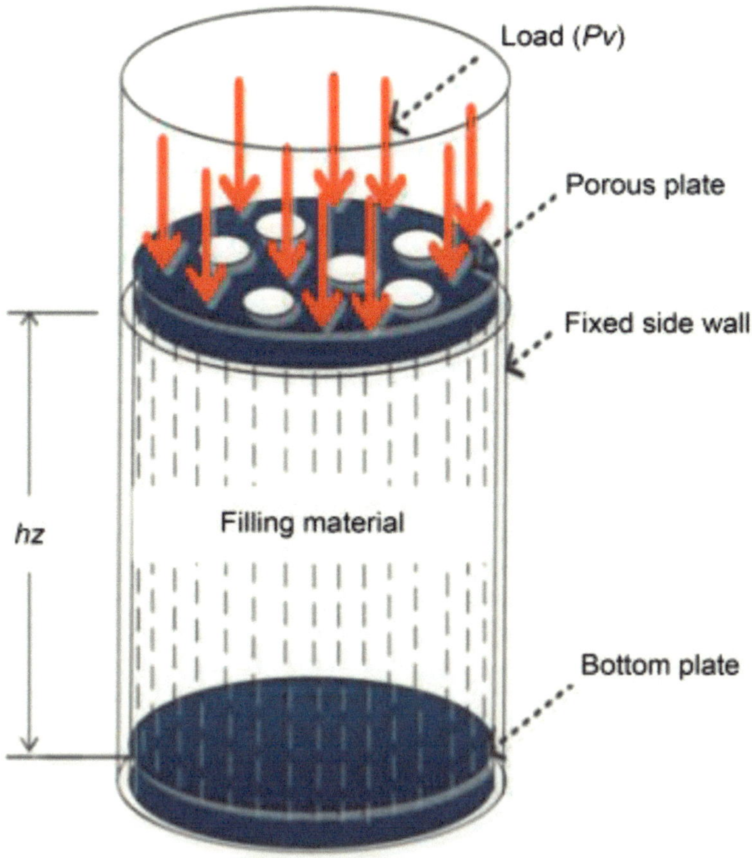

Figure 12.
Skeleton pore filled with filler.

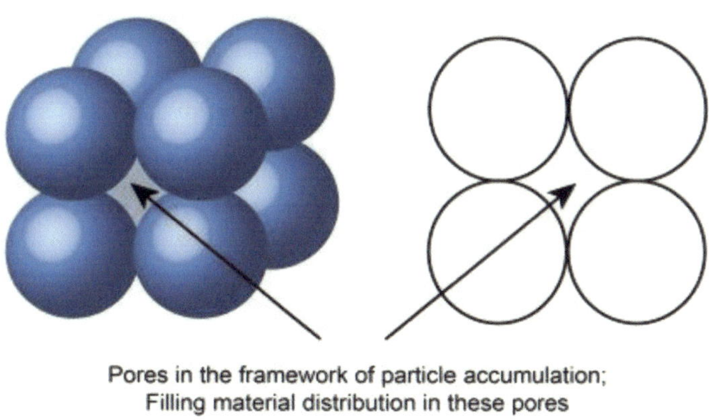

Figure 13.
Mixed material partial filling.

$$\Delta V_{mixed\ material} = \Delta V_{skeleton\ pore} = \frac{\Delta V_{filler}}{\beta} + \frac{V_{filler}}{\beta} - \frac{V_{filler}}{\beta_0} \tag{22}$$

When the filler grains expanded, the porous plate moved upward, which increased the macroscopic volume. In addition, some fine grains were extruded via the top plate pore, and plastic deformation consumed energy, as shown by formula (23).

$$\Delta V_{filler} = \Delta V_{mixed\ material} + \Delta V_{extrude} \tag{23}$$

Based on formulas (22) and (23),

$$\Delta V_{extrude} = \Delta V_{filler} - \frac{\Delta V_{filler}}{\beta} - \frac{V_{filler}}{\beta} + \frac{V_{filler}}{\beta_0} \tag{24}$$

Because filler grains are in a plastic state, their elastic potential energies cannot be increased. Therefore, $\Delta E_{elastic\ potential\ energy} = 0$.

The mixed material expands by ΔV_{all}, the upper load is N, and the pressure is $p_v = N/S_2$. Assume that the piston area is $S = S_1 + S_2$, where S_1 is the piston area with pores, and S_2 is the piston area without pores. Therefore, the work during expansion is

$$W_{work} = \Delta V_{all}\, p_v \tag{25}$$

Note: $\Delta V_{all}\, p_v$ has units of energy, whereas $\Delta V_{all}\, N$ does not have units of energy.

The extrusion volume is calculated via formula (24). The corresponding soil develops plastic deformation and dissipates energy. Based on a general expression of the plastic strain flow rule under the principle of minimum energy consumption, the plastic deformation energy consumption rate is

$$\psi = \int_v \rho\varphi dv \tag{26}$$

Where $\rho\varphi = \sigma d\varepsilon^p$. The energy dissipation process only occurs when the yield criterion is satisfied. In this paper, plastic energy is the product of the stress friction coefficient and volume strain. That is, $W_{plastic}$ deformation energy consumption is proportional to the soil internal friction coefficient K, soil stress P and soil extrusion volume $\Delta V_{extrude}$. In addition, when β is larger and the area of pores in the porous plate is smaller, it is more difficult to extrude soil, and energy consumption is higher. Therefore, it should be positively correlated with $1/(1-\beta)$. Plastic energy consumption is described by formula (27):

$$W_{plastic\ deformation\ energy\ consumption} = \frac{\chi K P \Delta V_{extrude}}{(1-\beta)} \tag{27}$$

where K is a dimensionless variable, P is stress, χ is a constant, and $\chi = (1-\beta_0)b$.

The solution for P now follows.

A microelement under a porous plate hole to be extruded is in a plastic flow state, and $\sigma_3 = 0$. $\sigma_1 = \sigma_2 > 0$. Based on the Mohr-Coulomb criterion, $\sigma_1 = \sigma_2 = 2c\cos\varphi/(1-\sin\varphi)$.

Therefore,

$$P = \frac{\sigma_1 + \sigma_2 + \sigma_3}{3} = \frac{4}{3}c\frac{\cos\varphi}{1-\sin\varphi} \tag{28}$$

K represents soil friction, and $K = \tan\varphi$.

In this process, the expansion potential energy is

$$\begin{aligned}\Delta H &= \Delta E_{elastic\ potential\ energy} + W_{work} + W_{plastic\ deformation\ energy\ consumption} \\ &= \Delta V_{mixed\ material} p_v + \chi K P \Delta V_{extrude} \frac{1}{1\text{-}\beta}\end{aligned} \tag{29}$$

Based on formulas (22), (24) and (29),

$$\begin{aligned}\Delta H = &\left(\frac{\Delta V_{mixed\ material}}{\beta} + \frac{V_{mixed\ material}}{\beta} - \frac{V_{mixed\ material}}{\beta_0}\right) p_v \\ &+ \chi K P \left(\Delta V_{mixed\ material} - \frac{\Delta V_{mixed\ material}}{\beta} - \frac{V_{mixed\ material}}{\beta} + \frac{V_{mixed\ material}}{\beta_0}\right) \frac{1}{1\text{-}\beta}\end{aligned} \tag{30}$$

According to the principle of minimum energy, the total internal energy of a closed system with steady volume, external parameters and entropy will tend to decrease. When a balance state is reached, the overall internal energy reaches the minimum level. β should result in a minimum ΔH; the range of β is $\beta_0 \le \beta \le 1$.

The prerequisite for minimum ΔH is

$$\frac{\partial \Delta H}{\partial \beta} = 0 \tag{31}$$

The filler grain volume expansion rate is $\alpha = \Delta V_{\text{mixed material}}/V_{\text{mixed material}}$, and $b = p_v/KP$; then, based on (31),

$$\left(\frac{1}{1-\beta}\chi - b\right)\left(\frac{\alpha}{\beta^2} + \frac{1}{\beta^2}\right) + \chi\left(\alpha - \frac{\alpha}{\beta} - \frac{1}{\beta} + \frac{1}{\beta_0}\right)\frac{1}{(1-\beta)^2} = 0 \tag{32}$$

When $\alpha = 0$, $\beta = \beta_0$. This is substituted into the above formula, yielding $\frac{1}{1-\beta_0}\chi - b = 0$, where $\chi = (1-\beta_0)b$.

It is easy to prove that when $\alpha \to +\infty$, $\beta \to 1^-$. The solution is

$$\left(\alpha\chi - b - \alpha b + \frac{\chi}{\beta_0}\right)\beta^2 + (2\alpha b - 2\chi - 2\alpha\chi + 2b)\beta + (\alpha\chi - b - \alpha b + \chi) = 0 \tag{33}$$

Because the range for β is $\beta_0 \le \beta \le 1$, the solution for the single variable quadratic equation (33) is

$$\beta = \frac{A \pm \sqrt{B - C \cdot D}}{E} \tag{34}$$

where

$$\begin{aligned} A &= -(2ab - 2\chi - 2\alpha\chi + 2b), \\ B &= (2ab - 2\chi - 2\alpha\chi + 2b)^2, \\ C &= 4\left(\alpha\chi - b - \alpha b + \frac{\chi}{\beta_0}\right), \\ D &= (\alpha\chi - b - \alpha b + \chi), \\ E &= 2\left(\alpha\chi - b - \alpha b + \frac{\chi}{\beta_0}\right). \end{aligned} \tag{35}$$

Based on formula (34), this is simplified to

$$\beta = \beta(\alpha, b, \chi, \beta_0) \tag{36}$$

where $b = p_v/KP$, $p_v = N/S = N(1+e)$, and $\chi = (1-\beta_0)b$.
Formula (36) is further optimized:

$$\beta = \beta(N, \alpha, C, \varphi, \beta_0) \tag{37}$$

The formula for calculating the partial filling model expansion filling rate shows that the expansion filling rate is closely related to parameters such as the overlying load, filler cohesion, friction angle and initial fill rate.

The relations between these physical variables are

$$e_{skeleton0} = \frac{V_{skeleton\ pore}}{V_{skeleton\ \mathrm{solid}}} \tag{38}$$

$$\beta_0 V_{skeleton\ pore} = V_{filler} \tag{39}$$

During expansion,

$$\begin{aligned} e_{skeleton} &= \frac{V_{skeleton\ pore} + \Delta V_{skeleton\ pore}}{V_{skeleton\ \mathrm{solid}}} = e_{skeleton0} + \frac{\Delta V_{skeleton\ pore}}{V_{skeleton\ \mathrm{solid}}} \\ &= e_0 + \frac{\Delta V_{filler}}{V_{skeleton\ \mathrm{solid}}}\left(\frac{\alpha}{\beta} + \frac{1}{\beta} - \frac{1}{\beta_0}\right) \end{aligned} \tag{40}$$

As the skeleton grain pore increases, $\Delta V_{skeleton\ pore} > 0$, $e_{skeleton}$ increases accordingly.

Based on formula (40), if there is only an elevation effect, $\beta_0 = \beta$, the ratio of $\Delta V_{skeleton\ pore}$ versus ΔV_{filler} does not change, and the following always holds:

$$V_{skeleton\ pore} + \Delta V_{skeleton\ pore} = \frac{(V_{filler} + \Delta V_{filler})}{\beta_0} \tag{41}$$

Ease of extrusion is directly related to β. In formula (40), if ΔV_{filler} is fixed and only β increases, then $\Delta V_{skeleton\ pore}$ decreases; i.e., small extrusion holes and lower elevations make extrusion difficult.

During filler grain expansion, elevation is calculated via formula (31), and extrusion is calculated via formula (32). The ratio of the two is calculated by substituting β into the two formulae.

5.3 Creation of a microheave filler frost heave model

In an actual project, when the filler is expanding, microheave filler skeleton pores cannot be filled completely. Therefore, the microheave filler frost heave complete filling model should be modified. The microheave filler complete filling theoretical model and partial filling model are combined, and interactions between the filling and elevation effects are considered. The model is created using the principle of minimum energy. For complete filling, the microheave filler frost heave model is described by formula (3). For partial filling, the microheave filler model is described by formula (23).

$$\Delta V_{mixed\ material} = \Delta V_{filler} - \left(\frac{\alpha - \frac{\alpha}{\beta} - \frac{1}{\beta} + \frac{1}{\beta_0}}{\frac{1}{\beta} - 1} \right) \left(V_{skeleton\ pore} - V_{filler} \right) \tag{42}$$

where

$$M = \frac{\alpha - \frac{\alpha}{\beta} - \frac{1}{\beta} + \frac{1}{\beta_0}}{\frac{1}{\beta} - 1} \tag{43}$$

The microheave filler frost heave model is simplified to

$$\Delta V_{mixed\ material} = \Delta V_{filler} - M \left(V_{skeleton\ pore} - V_{filler} \right) \tag{44}$$

The above formula is quantified. Based on formulas (4) and (44),

$$\begin{aligned} \Delta V_{mixed\ material} &= \Delta V_{filler} - M V_{residual\ pore} \\ &= \frac{m_{filler}}{\rho_{filler}} (M + \alpha) + M \frac{m_{keleton}}{\rho_{keleton}} - M V_{mixed\ material} \end{aligned} \tag{45}$$

Based on formulas (3), (45) is rearranged to

$$\Delta V_{混合料} = m_{骨架} \left[\frac{\frac{\theta}{1-\theta} (M + \alpha)}{\rho_{充填料}} + \frac{M}{\rho_{骨架}} \right] - M V_{混合料} \tag{46}$$

Based on formulas (7), (8) and (46),

$$\eta_{mixed\ material} = \frac{\Delta V_{mixed\ material}}{V_{mixed\ material}} = m_{skeleton} \left[\frac{\frac{\theta}{(1-\theta)} (M + \alpha)}{\rho_{filler}} + \frac{M}{\rho_{skeleton}} \right] / V_{mixed\ material} - M \tag{47}$$

Substitution of formulas (11) into (47) yields

$$\begin{aligned} \eta_{mixed\ material} &= m_{skeleton} \left[\frac{\frac{\theta}{(1-\theta)} (M + \alpha)}{\rho_{filler}} + \frac{M}{\rho_{skeleton}} \right] / V_{mixed\ material} - M \\ &= \rho_{mixed} (1 - \theta) \left[\frac{\frac{\theta}{1-\theta} (M + \alpha)}{\rho_{filler}} + \frac{M}{\rho_{skeleton}} \right] - M \end{aligned} \tag{48}$$

Based on the skeleton grain ratio definition, (49) is rearranged as

$$\begin{aligned} \eta_{mixed\ material} &= \rho_{mixed} (1 - \theta) \left[\frac{\frac{\theta}{1-\theta} (M + \alpha)}{\rho_{filler}} + \frac{M}{\rho_{skeleton}} \right] - M \\ &= \rho_{mixed} (1 - \theta) \left[\frac{\frac{\theta}{1-\theta} (M + \alpha)}{\rho_{filler}} + \frac{M}{G_{skeleton} \rho_w} \right] - M \end{aligned} \tag{49}$$

Based on formula (17), the left side is $\Delta V_{all} = \eta_{mixed}\, hS$. Based on the complete filling model, hS is 1. Rearranging formula (50) yields

$$\eta_{mixed\ material} = \rho_{mixed}(1-\theta)\left[\frac{\frac{\theta}{1-\theta}(M+\alpha)}{\rho_{filler}} + \frac{M}{\rho_{skeleton}}\right] - M$$
$$= \left[\frac{\rho_{mixed}(M+\alpha)}{d_s\rho_{filler}(1+0.01w)} + \frac{\rho_{mixed}(1-\theta)\alpha}{G_{skeleton}\rho_w}\right] - M \qquad (50)$$

where M is given by formula (43), and β in M is given by formula (35).

To determine β, $b = p_v/KP$ is used, where K represents soil friction, $K = \tan\varphi$, and P is determined using formula (33).

The pressure is $p_v = N/S = N(1 + e)$, and χ is a constant, where $\chi = (1\text{-}\beta_0)b$. Formula (51) is simplified to

$$\eta_{mixed\ material} = \eta(N, \varphi, C, \alpha, \beta_0, d_s, \rho_w, w, \rho_{mixed}, \theta, G_{skeleton}) \qquad (51)$$

The parameters in relation formula (51) were obtained via the indoor filler frost heave test, and the simple physical and mechanical test was used for the microheave filler.

Based on the microheave filler frost heave model, microheave filler frost heave is closely related to the specimen overlying load, filler plastic deformation friction angle, filler cohesion, filler frost heave rate, mixed material skeleton pore filling rate by the initial filler, mixed material ratio, water density, mixed material water content, mixed material density, filler content and mixed material skeleton ratio.

6. Model verification

Based on the model presented in this paper, the filler frost heave rates for various filler filling rates are shown in **Figure 14**, and the filler frost heave rates for various water contents are shown in **Figure 15**. These results were then compared with the test results presented in this paper.

Figures 14 and **15** show that the model simulation results essentially matched the test results. However, the specimens in the test and simulation differed. A silt filler frost heave rate test under an overlying load of N_1 = 60 kPa was performed. Other parameters, including the model parameters, are listed in **Table 5**.

The filler frost heave rate in the lab is $\eta_{mixed\ material}$ = 0.123 mm/ 150 mm = 0.082%.

The simulation showed that when the water content was 9%, the frost heave rate was approximately 0.22%.

A comparison of the calculated and experimental filler frost heave rates showed a difference of only 0.06%, which indicates that the microheave filler frost heave computational model derived in this paper is viable for frost heave rate calculations.

7. Conclusions

To solve problems of seasonal permafrost region subgrade filler microheave in China, the effects of influencing factors, such as the filler content, natural water holding capacity, filler grading, permeability coefficient and water supply

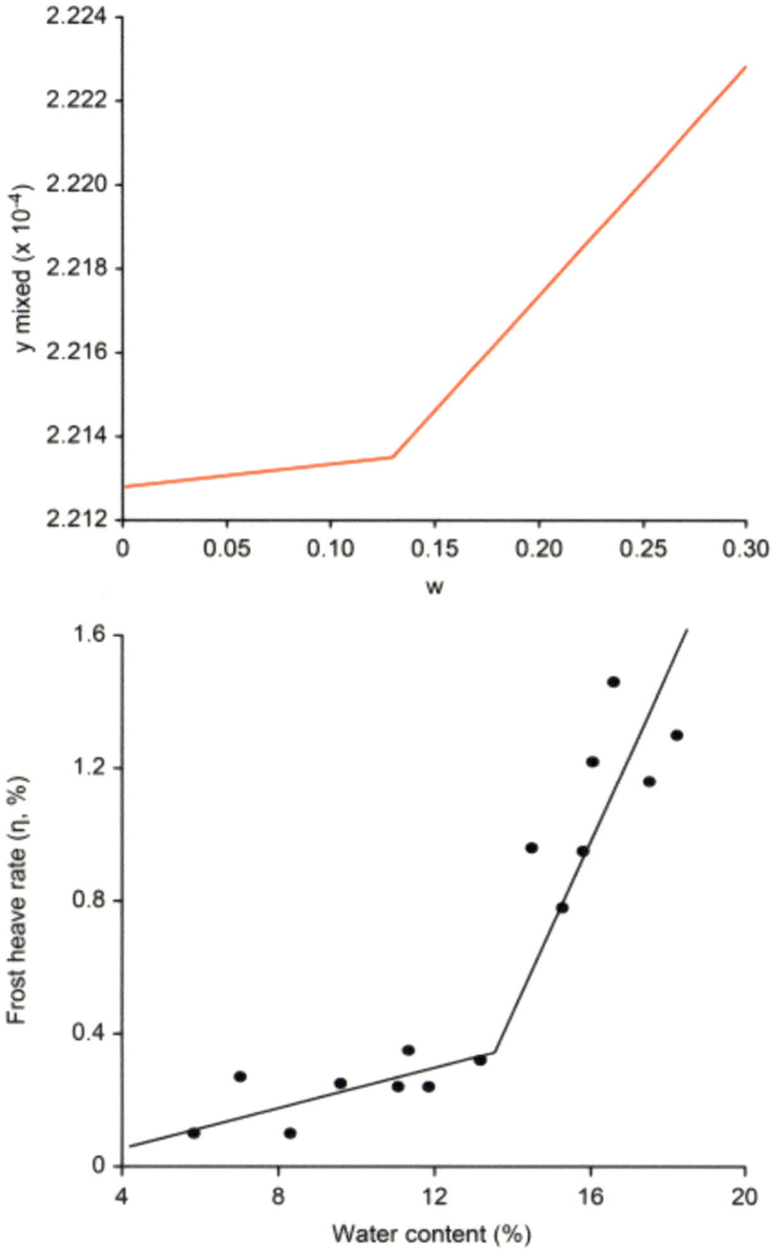

Figure 14.
Frost heave rate versus filler content.

condition, on the subgrade frost heave rate with or without external constraints (overlying train loads) were investigated via an indoor frost heave characteristics test. On that basis, models for complete and partial filling microheave filler frost heave were created. The validity of the theoretical microheave filler frost heave rate forecast model was demonstrated via indoor testing. The detailed conclusions are as follows.

1. The filler grading frost heave test showed that when the gravel or finer aggregate was reduced, the filler frost heave rate decreased gradually; however, even for very low levels of fine grain content at low temperatures,

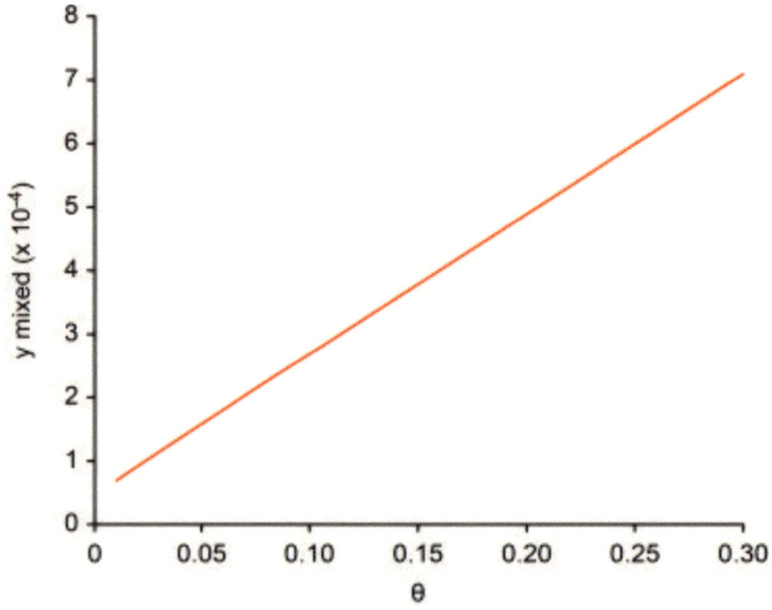

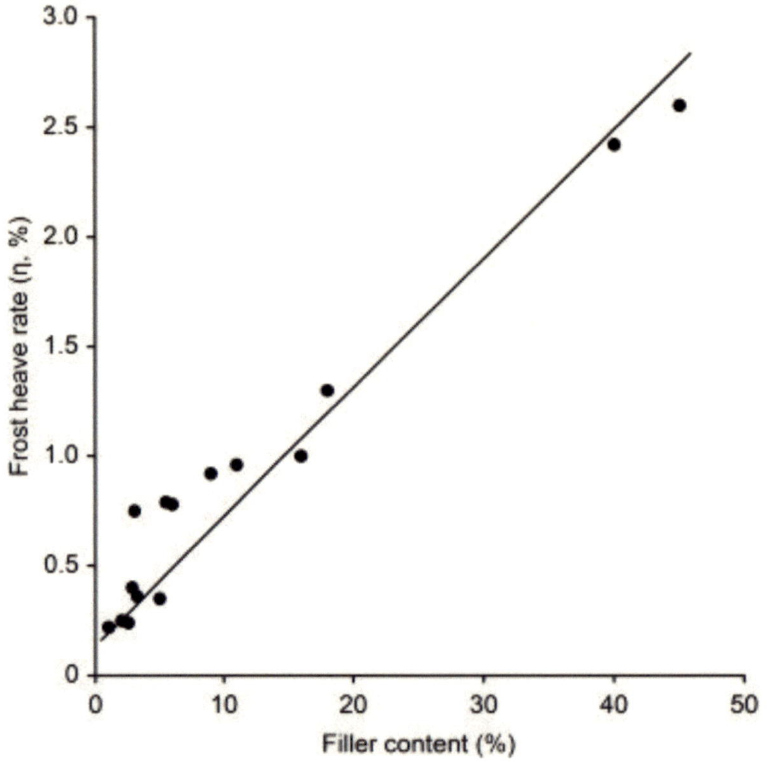

Figure 15.
Variations in frost heave rate with water content.

w_p	C_1	φ	ρ	e	β_0	w	α
9	6	30°	2.1	0.3	0.178	9%	0.22%

Table 5.
Model parameters.

the soil still produced a certain level of frost heave. The filler overlying load test showed that an overlying pressure produced a certain level of suppression effects on filler frost heave.

2. Microheave filler frost heave increased with filler frost heave; however, when the filler fill rate was under 0.25, as the filler content increased, the changes in filler frost heave were insignificant. This means that at this condition, filler frost heave filled pores, no filler macroscopic frost heave occurred, and soil was insensitive to frost heave. For filler filling rates exceeds 0.25, as the filling rate increased, the filler frost heave rate increased significantly, and the soil was sensitive to frost heave.

3. A microheave filler frost heave model, $\eta_{mixed\ material} = \eta(N, \varphi, C, \alpha, \beta_0, d_s, \rho_w, w, \rho_{mixed}, \theta, G_{skeleton})$, which considers the overlying load, filler plastic deformation friction angle, filler cohesion, filler frost heave rate, mixed material skeleton pore filling rate by the initial filler, mixed material ratio, water density, mixed material water content, mixed material density, filler content and mixed material skeleton ratio, was created. The applicability of the microheave filler frost heave model was verified via testing.

Acknowledgements

This study was sponsored by the Foundation of China Academy of Railway Sciences for the Youth (No.2017YJ050).

Notes

α	filler expansion rate
β	mixed material skeleton grain pore filling rate by filler grains
β_0	initial filling rate by filler grains
d_s	mixed material ratio
$\Delta E_{elastic\ potential\ energy}$	D-value of elastic potential energy
$G_{skeleton}$	skeleton grain ratio
Δh	overall specimen frost heave in mm
ΔH	expansion potential energy
H_f	frozen depth (excluding frost heave) in mm
η	microheave filler frost heave rate
$\eta_{mixed\ material}$	mixed material frost heave rate
η_{filler}	filler compactness
θ	filler content
K	soil internal friction coefficient, $K = \tan\varphi$
M_1	skeleton grain mass for a unit specimen volume
M_2	filler mass for a unit specimen volume
m_{filler}	quality of filler
$m_{mixed\ material}$	quality of mixed material
$m_{skeleton}$	quality of skeleton
N	upper load
ρ_{filler}	density of filler
$\rho_{mixed\ material}$	density of mixed material
$\rho_{skeleton}$	density of skeleton

ρ_w	water compactness
P	soil stress
p_v	pressure
γ_1	skeleton grain dry density for the corresponding soil sample compactness
γ_2	filler dry density for the corresponding soil sample compactness
s	specimen's filler volume fill rate
S	piston area
S_1	piston area with pores
S_2	piston area without pores
V_1	filler skeleton grain volume fraction
V_2	filler volume fraction
$V_{mixed\ material}$	mixed material grain volume fraction
ΔV_{all}	mixed material expands
$\Delta V_{extrude}$	soil extrusion volume
ΔV_{filler}	D-value of filler volume fraction
$\Delta V_{mixed\ material}$	D-value of mixed material grain volume fraction
$\Delta V_{residual\ pore}$	D-value of residual pore volume fraction
$\Delta V_{skeleton\ pore}$	D-value of skeleton pore volume fraction
ω	difference between the filler initial water content
ω_0	filler optimal water content
ω_p	filler plastic limit water content
ω_v	filler volumetric water content
$W_{plastic\ deformation\ energy\ consumption}$	work during plastic deformation energy consumption
W_{work}	work during expansion
w	microheave filler water content
ψ	plastic deformation energy consumption rate
χ	a constant

Author details

Ye Yangsheng, Du Xiaoyan, Zhang Qianli and Chai Jinfei*
China Academy of Railway Sciences, Railway Engineering Research Institute, Beijing, China

*Address all correspondence to: chaijinfei@gmail.com

References

[1] Ye YS, Cai DG, Yan HY, Yao JP, Zhang QL, Cheng AJ. The mechanism analysis of micro frost heave coarse-grained filling material in the high-speed railway. Japanese Geotechnical Society Special Publication. 2016;**2**(47): 1660-1663

[2] Chen X, Liu J, Liu H. Frost action of soil and foundation engineering. Beijing: Science Press; 2011 (in Chinese)

[3] Askar Z, Zhanbolat S. Experimental investigations of freezing soils at ground conditions of Astana, Kazakhstan. Sciences in Cold and Arid Regions. 2015; **7**(4):0399-0406

[4] Konrad J-M. Freezing-induced water migration in compacted base-course materials. Canadian Geotechnical Journal. 2008;**45**(7):895-909

[5] Konrad JM, Lemieux N. Influence of fines on frost heave characteristics of a well-graded base-course material. Canadian Geotechnical Journal. 2005; **42**(2):515-527

[6] Bilodeau J-P, Dore G, Pierre P. Gradation influence on frost susceptibility of base granular materials. International Journal of Pavement Engineering. 2008;**9**(6):397-411

[7] Uthus L, Hermansson Å, Horvli I, Hoff I. A study on the influence of water and fines on the deformation properties and frost heave of unbound aggregates. In: International Conference on Cold Regions Engineering, pp. 1-13

[8] Wang T, Yue Z, Sun T. Influence of fines content on the anti-frost properties of coarse-grained soil. Sciences in Cold and Arid Regions. 2015;**7**(4):0407-0413

[9] Bilodeau JP, Doré G, Poupart J. Permanent deformation of various unbound aggregates submitted to seasonal frost conditions. Cold Regions Engineering. 2012:155-164

[10] Guo Y, Yang Y, Zeng L. Study on frost heaving characteristics of subgrade soil of lanzhou-xinjiang railway. Subgrade Engineering. 2010;**2**:149-151

[11] Sheng DC, Zhang S, Niu F, Cheng G. A potential new frost heave mechanism in high-speed railway embankments. Géotechnique. 2014;**64**(2):144-154

[12] Abdalla B, Fan C, Mckinnon C, Gaffard V, Audibert-Hayet A, Coche E, et al. Extended porosity rate function for frost heave. In: ASME 2014 33rd International Conference on Ocean, Offshore and Arctic Engineering, ASME. San Francisco, CA; 2010. V010T007A046

[13] Zhang Y, Michalowski RL. Multiphysical modeling and numerical simulation of frost heave and thaw settlement. In: Geo-congress 2014 Technical papers@ sgeo-characterization and modeling for sustainability, ASCE, Atlanta, GA; 2014. pp. 2735-2744

[14] Bronfenbrener L. The modelling of the freezing process in fine-grained porous media: Application to the frost heave estimation. Cold Regions Science and Technology. 2009;**56**(2):120-134

[15] Yin C. Study on Seasonal Permafrost Region Subgrade soil Frost Heave Characteristics and Evaluation Index System. Northeast Forestry University; 2014

[16] Wang Z, Zhang Q, Ye Y. In-depth study on the classification of filling for railway subgrade. China Railway Science. 2012;**33**(2):13-20

[17] Li A, Niu YH, Niu F, Liu H. Research status of frost heaving

properties and controlling measures of coarse grained soil. Journal of Glaciology and Geocryology. 2015;**37**(1):202-210

[18] Du X. Research on Filler Frost Heave and Coarse Grain Skeleton Interaction-Based Micro Heave Filler Frost Heave Mechanism. Beijing: China Academy of Railway Sciences; 2015

[19] Tian Y, Liu J, Peng L. Experimental study on frost action of fine-grained soils under dynamic and static loads. Chinese Journal of Geotechnical Engineering. 2010;**32**(12):1882-1888

[20] Zhao H, Yan H, Zhang Q, et al. Study on frost-heave performances of filler consisting of coarse grained soils for subgrade bed in seasonal frozen region. Railway Engineering. 2014;**7**: 92-94

Chapter 6

Estimation of Excess Pore Pressure Generation and Nonlinear Site Response of Liquefied Areas

Kamil Bekir Afacan

Abstract

Recent studies about liquefaction initiation are widely encountered in the literature in terms of utilizing the dynamic triaxial tests under harmonic loading and site response of liquefied zones. Sandy-like or clayey-like behavior is important for estimating the liquefaction susceptibility but there are other factors related to cyclic loading characteristics such as frequency content and stress level. Besides, 1-D ground response analyses are employed to understand the behavioral transmission through the soil column in liquefiable areas. The study here focuses on two main aspects of the liquefaction. The first part consists of the estimating of the pore pressure generation under irregular excitations, whereas the second part aims to assess the efficiency of the building codes predicting the nonlinear site response in liquefied prone areas. The laboratory results show that the frequency content has big influence on the liquefaction at varying stress levels. Moreover, literature models have discrepancies to estimate the pore pressure generation under different types of loading. Regarding the site response, it was indicated that equivalent linear approach is incapable of predicting the seismic behavior of soil column; therefore, nonlinear ground response must be run instead, and the IBC is the most effective one to match the nonlinear analysis results.

Keywords: liquefaction, pore pressure generation, nonlinear site response, dynamic soil behavior, building codes

1. Introduction

Liquefaction is one of the most important and complex topics that geotechnical engineers deal with in today's practice. The Good Friday (Alaska) and Niigata (Japan) earthquakes in 1964 are the very first earthquakes that the destructive effect of the liquefaction was observed as the slope failures, bearing failures of foundations and abutments, etc. It has been studied widely since, and so many researchers have helped liquefaction to be understood better although the topic is still complicated in many sense. The initiation of the studies starts in the 1960s by Seed and his colleagues. Seed and Ldriss in 1967 [1], Seed in 1968 [2], and Seed and Peacock in 1971 [3] set the base of the liquefaction concept by studying the clean sand samples to evaluate the cyclic resistance of the soils, and Finn et al. [4, 5], Castro [6, 7], and Youd [8, 9] are other examples that focused on the liquefaction around that time.

Cyclic loading of the saturated coarse soils causes excessive pore pressure buildup under undrained conditions. This excessive pore pressure can converge up to the total stress; meaning zero effective stress is described as liquefaction [10]. It means that the soil is no longer capable of carrying any load resulting in soil failures and/or excessive deformations of structure foundations. Therefore, this phenomenon is very crucial to take into consideration in designing a structure in liquefaction susceptible areas.

The process of the pore pressure buildup mainly depends on the wave propagation throughout the soil column that is excited by an earthquake. Loose soils tend to densify when they are loaded under drained conditions, whereas if the condition happens to be undrained, the compression of the soil during cyclic loading is almost impossible. In this case, there is not enough time for water to escape from the voids; therefore it takes the seismic load reflecting as increase in the pore pressure. This pore pressure starts building up until the loading is finished. If the energy created by the excitation is big enough, the amount of the pressure will finally reach to the total stress. Since there is no effective stress anymore, the soil grains will escalate in the water and lose its strength therefore stability. Thus, the liquefaction susceptibility should be questioned in seismic areas with some basics: Is the soil coarse and loose enough for being susceptible to liquefy? Is the region active enough to produce sufficient energy to liquefy the soil? If so, does the liquefaction happen in shallow depths that will greatly affect the superstructure on it, or does it happen very deep that nothing will be felt on the ground? These are the questions that should be answered in seismically active regions to take initial precautions for possible hazards.

Although the liquefaction susceptibility is correlated with the in situ test results (SPT, CPT, Vs) or soil index parameters (relative density, fine content, Atterberg limits, etc.) in practice, today's technology gives engineers big confidence to run laboratory cyclic tests in order to understand the behavior of the soil under seismic loads. It is understandable to use empirical models to evaluate the liquefaction; because of its easiness, they are not as accurate as the real scale pore pressure models observed by the dynamic triaxial tests conducted in the laboratory. The complexity of the excess pore pressure generation models is still there, but it is the most appropriate way to estimate the dynamic behavior of coarse soils. Since the correlations do not give any information about the pore pressure generation, the possible deformations on the ground due to liquefaction is not more than an approach for them; however, the level of deformation levels can be observed in the laboratory rigorously [11]. It should also be noted that the tests can be run in varied strain amplitudes (10^{-6}–10^{0}).

The cyclic behavior of soils is commonly determined by the dynamic triaxial test equipment in the laboratory. Pore pressure generations models are mainly studied into three different approaches: (i) stress-based models, (ii) strain-based models, and (iii) energy-based models. The purpose of stress-based models is to relate the excess pore pressure ratio (r_u) which is the normalization of the excess pore pressure by the effective stress with the number of cycles to liquefy the soil. Lee and Albaisa [12] and Seed et al. [13] studied the relation between the r_u with (N/N_{liq}) where N is the number of loading cycle and N_{liq} is the number of cycle to initiate liquefaction. Later, Booker et al. [14] worked in the same model but offered different constants, and Polito et al. [15] in 2008 came up with a model consisting of fine content, relative density, and cyclic stress ratio. Finally, Baziar et al. [16] updated the formulation with some new parameters. Regarding the strain-based models, the change in volumetric strain is estimated with the cyclic shear strain along with some constants for the clean sand samples [4, 5, 16, 17]. More complex models were introduced by Dobry et al. [18–20], Vucetic and Dobry [21], and Carlton [22] with new parameters such as threshold cyclic shear strain, the dimension of the loading and other fitting parameters in relation to fine content, shear wave velocity, etc. Energy-based models focus on the stress-strain loops to calculate the enough energy to estimate the liquefaction susceptibility, and Green et al. [23] is one the most widely used models in this concept. In this study, the laboratory tests were conducted at different cyclic stress ratios (CSR), and the results were compared with the stress-based models. The details of these models and their formulation will be discussed later.

As mentioned before, loss of stiffness of soil is a big problem for civil engineers. Kramer et al. [24] states that estimation of the soils behavior under dynamic loading is crucial to construct any kind of structures reliable and economical. Other than laboratory studies, building codes regulate the design for a structure to be built in a specific region to estimate the dynamic loads that will be transmitted throughout the soil layers to the structure for some seismic scenarios. Ground response varies a lot with the dynamic behavior of soils and local site conditions under seismic activity. The building codes generally construct the response spectrum differently with regard to the soil class which is determined by the shear wave velocity of upper 30 m. The classification of this kind misses the local site effects on the ground response that the structure will be exposed to. Therefore, the site-specific response plays an important role to estimate the dynamic loads propagated to the soil surface. Building codes generally set some rules to necessitate the site response (i.e., ZF soil class); however, the prediction of the dynamic loads that structure would feel will change drastically by the nonlinearity of soil behavior.

Dynamic behavior of soils is obtained in the laboratory conducting dynamic triaxial tests for medium to high and resonant columns test for low to medium shear strain levels. The behavior is modeled with the modulus reduction and damping properties of soil at varying shear strains. These values are evaluated from the stress-strain loops for every cycle to build two important curves presenting the soil nonlinearity.

This chapter will outline the liquefaction studies into two groups which will be (i) experimental and (ii) analytical approaches. Initiation of the excess pore pressure buildup resulting $r_u = 1$ ends up as liquefaction of soil, and it can mainly be assessed in the laboratory by the dynamic triaxial test systems. Eighteen different dynamic tests were conducted at different CSRs and loading types in order to obtain the pore pressure generation and validate the models in the literature with the recorded data. The second part of the study will focus on the nonlinear site response of the liquefied areas with the available in situ data. Five different boring logs were used in the analyses, and the nonlinear behavior of the soil layers was determined for each. It should be noted that all the soil columns were liquefied at different depths for different earthquake scenarios. The ground responses were calculated, and finally the performance of the building codes to estimate the spectral response of the soil columns was assessed.

2. Estimation of the excess pore pressure generation

In this part, clean sand samples were loaded under varying cyclic stresses with different loading patterns using the dynamic triaxial test system equipped at the Eskisehir Osmangazi University Soil Dynamics Laboratory, and it is presented in **Figure 1**.

The system is a fully automated pneumatic system that is capable of conducting axial strains up to 10%, and the tests can be handled as stress controlled or strain controlled. A total of 18 tests was completed to justify the performance of the excess pore pressure generation models suggested in the literature.

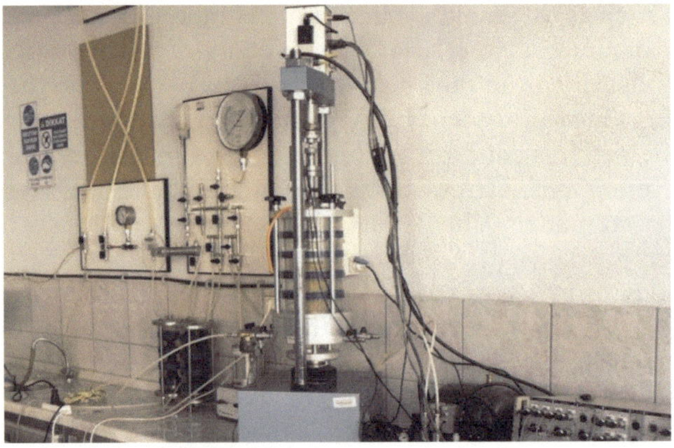

Figure 1.
Dynamic triaxial test system at the Eskisehir Osmangazi University.

2.1 Material and method

In order to study liquefaction in the laboratory, standard sand samples are generally chosen for the tests since the cost of the undisturbed sand samples that is only possible by the freezing technique are really high. Therefore, clean Podima sand samples were used in this study. The type of sand is classified as poorly graded sand (SP) according to the unified soil classification system (USCS), and the grain size distribution is shown in **Figure 2**. The minimum and maximum unit weights of the sand were determined as 15.30 and 17.66 kN/m^3 as the standards suggest, respectively. The specific gravity was determined as 2.63.

The soil samples had a diameter of 7 cm and a height of 14 cm, and the desired relative density was decided to be as 40%. Dried sand was dropped into a two parted mold, which has a membrane placed in, by a funnel with some little shaking and compacting of the sample accordingly. Later, the top header was set on the sample, and some vacuum was applied to the sample in order to hold itself, and the mold was taken apart. Following external cell placement, the cell was

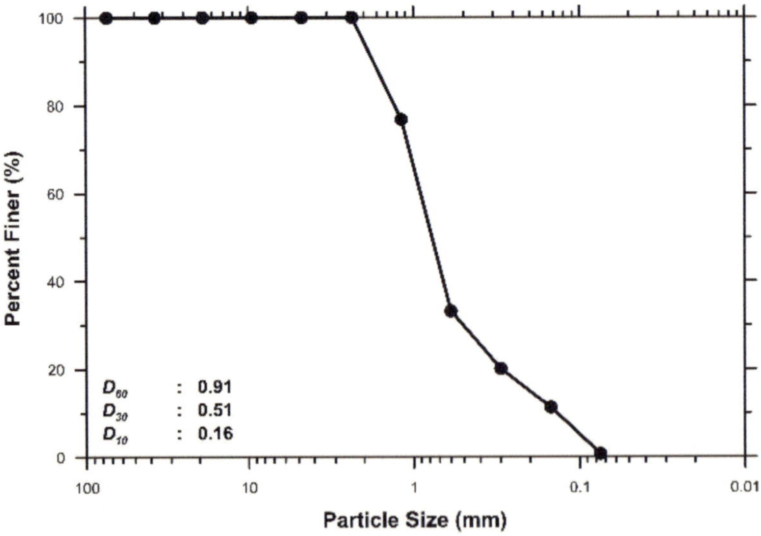

Figure 2.
Grain size distribution of the sand used in the study.

filled with water, and some cell and back pressures were applied in such a way that there was a difference of 5 kPa at increasing level. The saturation was controlled with the B constant to reach a value of 0.96. After saturation was accomplished, the sample was consolidated under isotropic condition approximately an hour with a confining pressure of 65 kPa. Finally, soil samples were axially loaded at different amplitude of stresses with changing loading shapes until all the samples were liquefied.

2.2 Test program and loading patterns

In order to understand the excess pore pressure generation at different cyclic stresses, three different stress levels were considered. Cyclic stress ratio is defined as the cyclic stress applied to the sample (either compression or extension) normalized by the confining pressure, and they were assigned as 0.15, 0.20, and 0.30 for the test program. All the samples were confined by the cell pressure of 65 kPa. The text program is shown in **Table 1**.

As seen in the table, a total of 18 tests was programmed to see the effect of the stress level and loading characteristics. It is customary to use harmonic loading in the liquefaction studies; however, irregular loading shapes were used in this study to investigate the differences. Six different patterns were used to axially load the samples, and the double amplitude of axial stresses was kept constant at each CSRs. The loading shapes at CSR = 0.20 level are presented in **Figure 3**.

2.3 Excess pore pressure generation models

Since the tests were assessed as stress controlled, the stress-based models were chosen to measure their performance with the laboratory results. Four different equations suggested by the models are presented in **Table 2**.

In the table, N is the specific cycle that the excess pore pressure ratio to be calculated and N_L is the number of cycles that liquefies the soil. α is a fitting constant, and although Seed and Booker propose as it is taken to be 0.7, Polito et al. advanced this constant with other parameters such as fine content (FC), relative density (D_r), and CSR. More complex model was offered by Baziar in 2011 with two different constants θ and β. These constants differ with the content of the silt in the sandy soil. Since clean sand samples were tested in this study, only related constants are shown in the table. The model Booker proposed gives almost identical results as Seed et al.'s model does; therefore it is not presented in the following part.

2.4 Observations

A total of 18 cyclic triaxial tests was planned to gather excess pore pressure generation at different stresses with changing excitation patterns. Using the necessary information from the test data, the excess pore pressure ratios (r_u) which were the ratio of the excess pore pressure recorded during cyclic loading over the

CSR	Double amp. axial str. (kPa)	Type 1	Type 2	Type 3	Type 4	Type 5	Type 6
0.15	20	X	X	X	X	X	X
0.20	26	X	X	X	X	X	X
0.30	39	X	X	X	X	X	X

Table 1.
Test program in terms of stress levels and loading types.

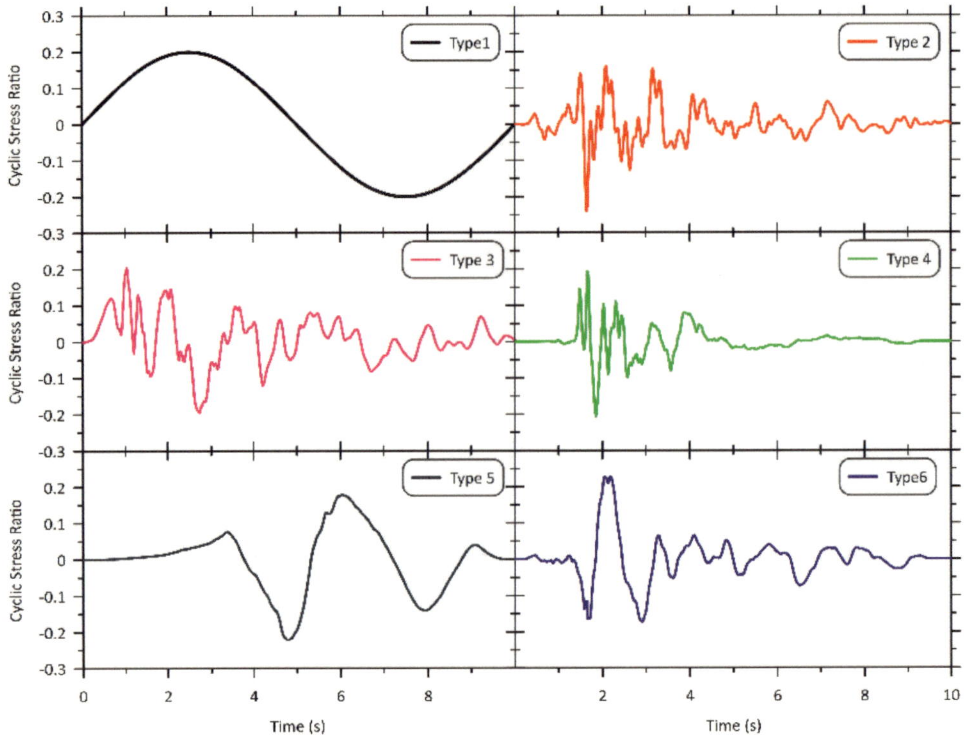

Figure 3.
Example loading shapes at the level of CSR = 0.20.

Models	Equation	Notes
Seed et al. [13]	$r_u = \frac{1}{2} + \frac{1}{\pi}\sin^{-1}\left[2\left(\frac{N}{N_L}\right)^{1/\alpha} - 1\right]$	$\alpha = 0.7$
Booker [14]	$r_u = \frac{2}{\pi}\sin^{-1}\left[\left(\frac{N}{N_L}\right)^{1/\alpha}\right]$	$\alpha = 0.7$
Polito et al. [15]	$r_u = \frac{2}{\pi}\sin^{-1}\left[\left(\frac{N}{N_L}\right)^{1/\alpha}\right]$	$\alpha = 0.01166 \cdot FC + 0.007397 \cdot D_r + 0.01034 \cdot CSR + 0.5058$
Baziar [16]	$r_u = \frac{2}{\pi}\sin^{-1}\left[\left(\frac{N}{N_L}\right)^{1/2\theta}\right] + \beta\sqrt{1 - \left\{\frac{2N}{N_L} - 1\right\}^2}$	$0.6 \le \theta \le 0.8$ $0.0 \le \beta \le 0.25$

Table 2.
Models and associated equations used to compare the test results.

confining pressure of 65 kPa were calculated for the models mentioned in the previous section. The test data and corresponding models proposed by Seed et al. [13], Polito et al. [15], and Baziar [16] at CSR = 0.15 are presented in **Figure 4**.

The first thing to point out in the figure is that the liquefaction happens in various times for different types of loading, although the double amplitude axial stresses are the same for all the six tests. The longest time to liquefy the clean sand sample happened to be almost 18,500 seconds for the sinusoidal type of loading, whereas this duration came out to be in an interval of 3500–8600 seconds for irregular type of loading. As seen in the upper left side of the figure, the recorded pore pressure buildup differs a lot with regard to other types. An initial increase was recorded in a short time, and it fluctuated for almost 10,000 seconds, and then the

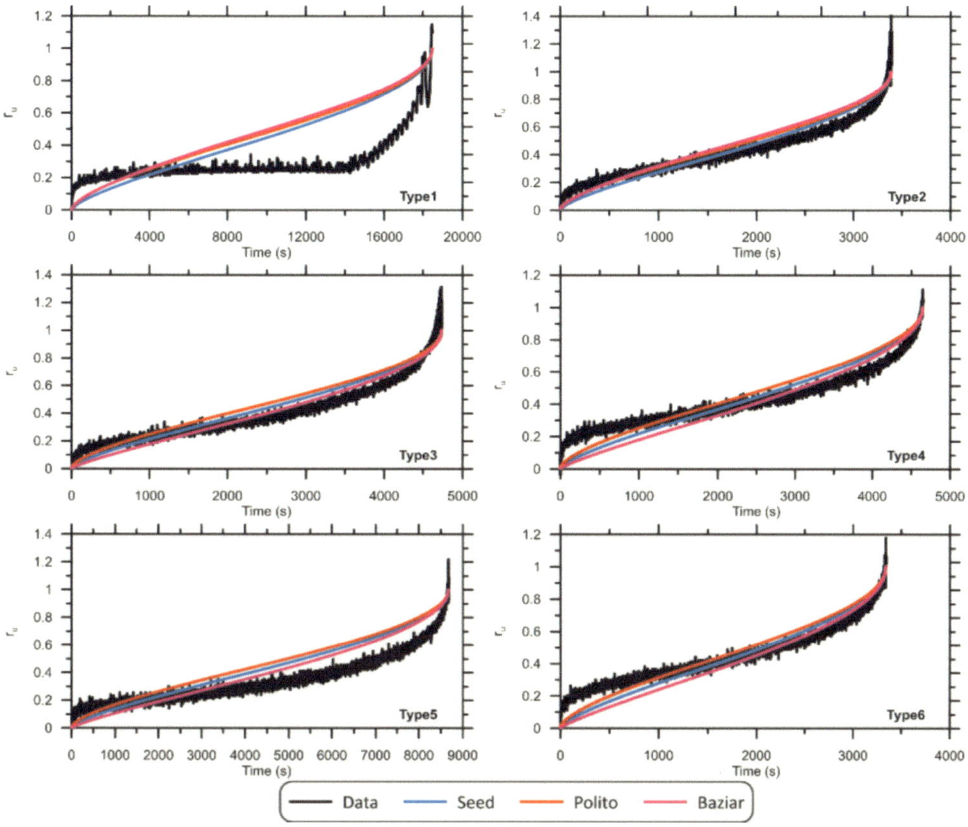

Figure 4.
Excess pore pressure ratios for the test data and models at CSR = 0.15.

accumulation of the excess pressure started to increase again, and then liquefaction finally occurred after 5000 seconds for Type 1. Similar trends were observed for the rest of the tests for shorter periods.

Regarding the models suggested in the literature, they do not behave much different at this stress level. Although they did not do a good estimation to represent the data gathered from the sinusoidal loading, other pore pressure ratios along time were predicted alright by the models, being slightly on the conservative side. It should be noted that all three models missed the initial increases in pore pressure regardless of type of loading. Lastly, the pore pressure observed from the type five loading acts more like a harmonic loading than irregular type of loading, and models overestimated the pore pressure generation after 2000 seconds until the liquefaction developed. Overall, some model has better presentation of the observed data at certain type of loading, and some does at others. Therefore, no model perfectly matches with the test data.

Moving on to the higher stress level, **Figure 5** shows the recorded excess pore pressure ratios at CSR = 0.20 for six different tests and corresponding estimations by the literature models. Compared to the lower stress level, the liquefaction initiation did not take that long under cyclic stress ratio of 0.20. It occurred around 130–420 seconds for soil to lose its stability for irregular types of loading, whereas it took almost 2.5–8 times longer for soil to liquefy under harmonic loading. The literature models ideally followed the pore pressure record gathered from the three tests under loading of Type 3, Type 5, and Type 6; Baziar, being the best, did acceptable predictions for other two tests (Type 2 and Type 4 loadings).

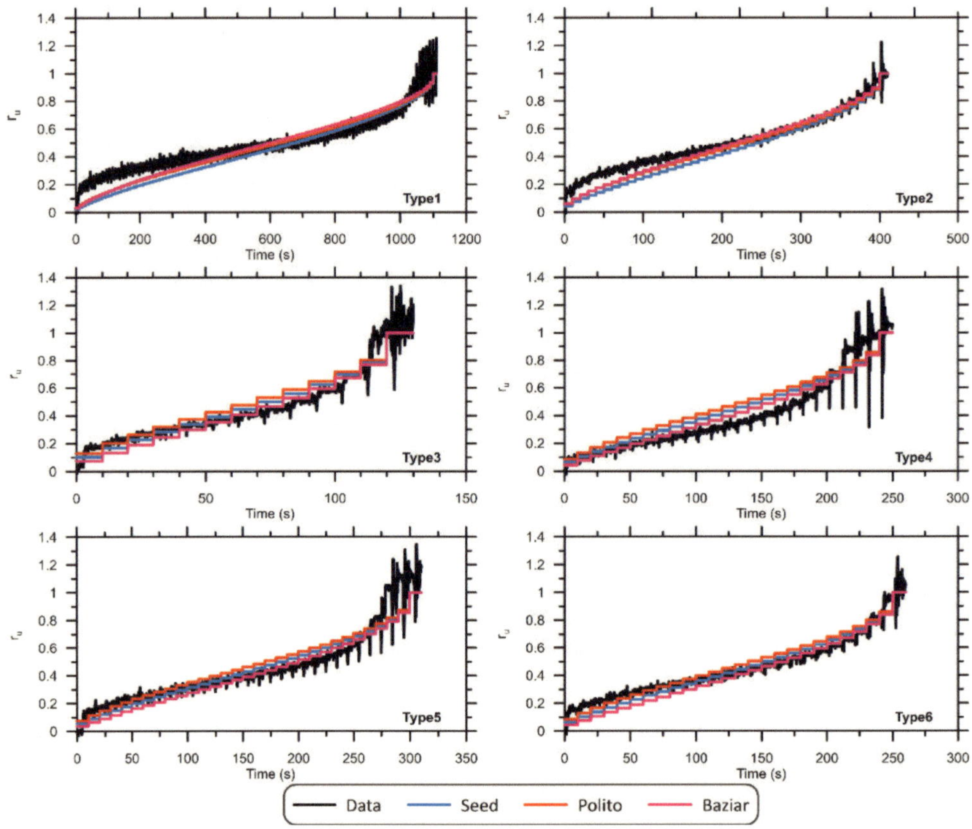

Figure 5.
Excess pore pressure ratios for the test data and models at CSR = 0.20.

As the axial stress increases, the pore pressure buildup behavior under sinusoidal loading converges to the one loaded by the irregular type. There is still a mismatch at the initial, but after 400 seconds, all three models presented the data alright for the Type 1 loading. It is obvious that most of the time the model proposed by Baziar [16] is the closest to the recorded data, and Polito's model [15] gives the upper limit out of the three models, whereas the other model stayed in between two.

Final set of tests were carried out at a high stress (CSR = 0.30) and recorded excess pore pressure ratios over time, and corresponding literature predictions are presented in **Figure 6**. It is obvious that the increasing stress level causes soil to liquefy at low number of cycles. For all the types of loading, it did take only 1–10 cycles for soil to lose its rigidity meaning not more than 100 seconds. As seen in the figure, all of the literature models predicted the pore pressure generation almost perfectly for the test done using Type 2 loading. Regarding the others, the data are underestimated by all the models especially the test data observed from the harmonic loading. It is hard to estimate the performance of the models to represent the data under Type 5 loading since it took only 1 cycle (10 seconds) to liquefy the soil.

There is one question that can arise to understand the behavior of the liquefaction occurred when harmonic loading was applied: Did the soil liquefy at 100 seconds or earlier like 30 seconds where the first time r_u value reached 1.0? During the specific test, the deformation of the soil was carefully observed with the naked eye along with the computer screen. Although r_u came to 1.0 around 30 seconds, the soil was not totally unstable and still was able to resist more loads. Therefore, the total stability lost was waited to terminate the test, and it lasted almost 100 seconds.

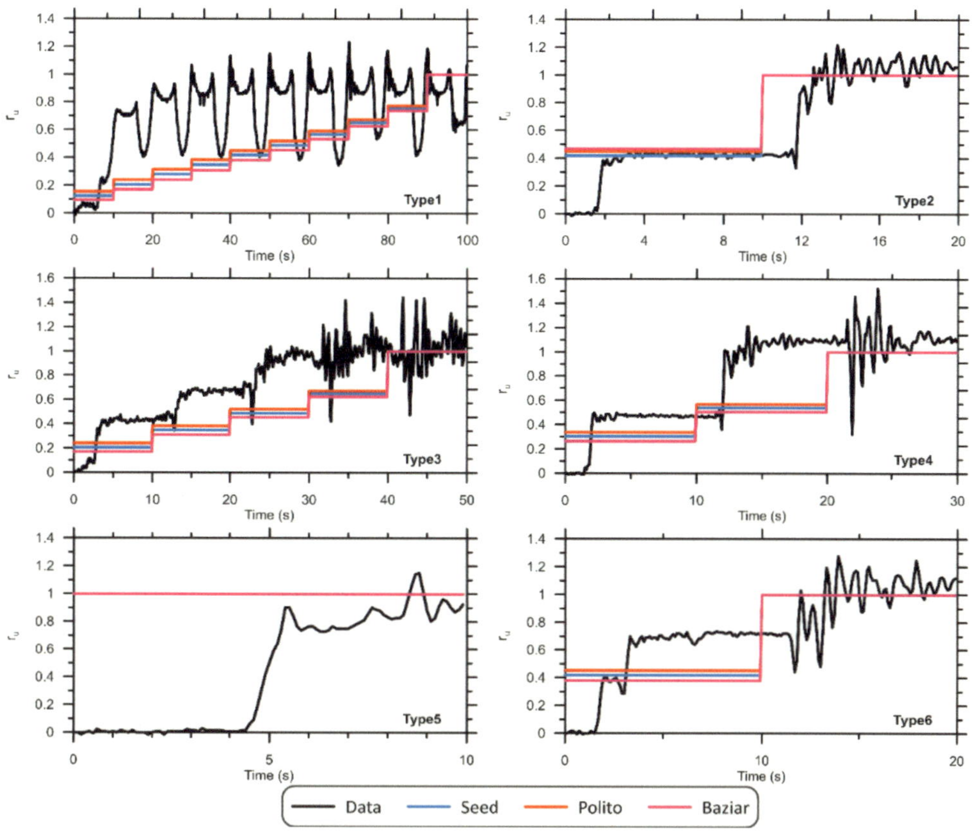

Figure 6.
Excess pore pressure ratios for the test data and models at CSR = 0.30.

2.5 Discussion

Out of 18 cyclic triaxial tests to understand the soil liquefaction susceptibility, it is observed that the shape of the loading or in other words frequency content of the excitation plays an important role to cause instability. Earthquakes are never harmonic types of excitations, and using sinusoidal loading in order to understand the soils dynamic behavior may mislead the interpretation. **Figure 7** explains how many

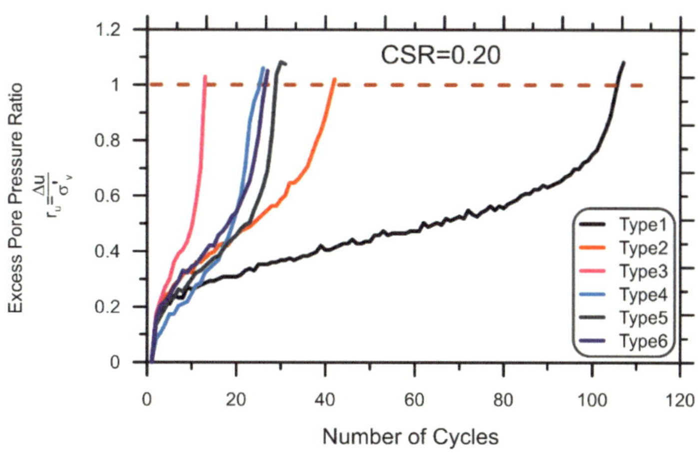

Figure 7.
The effect of loading on the change of number of cycles to liquefy soil.

cycles are needed to liquefy soil sample for different types of loading at a double amplitude of 26 kPa stress, just to give an example.

One clear point can be made from the figure that the variety of loading and its dominant frequency are effective on the number of cycles that would initiate the liquefaction. Although the excess pore pressure buildups are similar for the first 5 cycles, Type 1 with the harmonic shape diverges from the rest of the group, and the number of cycles needed for liquefaction occurs to be at least 2.5 times later than the others. With regard to the irregular type of loadings, they even act differently, and the number of cycles to liquefy the soil varies between 10 and 40 which can be considered a wide interval. Thus, the potential of liquefaction triggering should be studied with different types of loading when needed.

The essence of the results in terms of number of cycles to liquefy soil samples is shown in **Tables 3** and **4**.

It should be noted that the variation is a lot for different cyclic stress ratios. For example, Type 3 and Type 4 loading patters need similar number of cycles to liquefy the soil at CSR = 0.15, whereas it changes for increasing CSRs. Harmonic loading always takes more time/cycle to generate pore pressure than others, and every irregular loading type has its varying frequency-dependent characteristics at increasing stresses. Therefore, the excess pore pressure generation is not only affected by the frequency content of the loading alone, but also the stress levels along with it play important role in estimating liquefaction triggering.

As a summary, three different stress levels at varying loading shapes were used to run dynamic triaxial tests in order to determine the generation of excess pore pressures under cyclic excitations. The desired relative density and confining pressure for all 18 tests were constant as 40% and 65 kPa, respectively. There are a few clear points that are worth to emphasize:

i. The pattern of the loading (harmonic vs. inharmonic) is an effective parameter that influences the duration and the number of cycles to liquefy same soil samples.

ii. As the stress level increases, the number of cycles for soil to lose its stability decreases. For example, the needed time to liquefy soil at CSR = 0.15 changes

CSR	Type 1	Type 2	Type 3	Type 4	Type 5	Type 6
0.15	1848	340	474	466	867	335
0.20	111	41	13	25	31	26
0.30	10	2	5	3	1	2

Table 3.
Number of cycles to initiate liquefaction for different types of loading.

Boring	Depth (m)	V_s (m/s)	$V_{s,30}$ (m/s)	FC (%)
B1	20	150–280	220	10–85
B2	20	180–190	185	5–42
B3	20	165–290	230	27–35
B4	20	145–195	175	8–32
B5	20	165–175	170	18–38

Table 4.
Soil profiles.

to between 3500 and 18,500 seconds, and these values drop to 130–1100 seconds at CSR of 0.20. Even a small difference in the applied stress amplitude may cause a drastic drop in the number of cycles.

iii. There are many stress-based models to estimate the pore pressure buildup in the literature, and three of them were used here to see their performance to estimate the laboratory data. Models have better representation of the pore pressure generation at the low to medium stress levels for irregular loading patterns, and at high stress levels the capability to estimate the data is not that well. Among the models, the model Baziar et al. [16] suggests relatively does better jobs, and Polito et al.'s [15] model is being on the safer side all the time. However, the performance of all three to predict the recorded data does not differ that much.

iv. The dynamic triaxial tests take some time to set the sand sample appropriately and evaluate the data to understand the behavior, but other levels of relative densities and confining pressures should be tested to have better understanding of the pore pressure generation concept.

v. The latest point is that the effect of fine content should also be investigated, and if the opportunities (budget, facility, etc.) are possible, undisturbed samples should directly be taken out of the site to see if the real in situ samples are represented well by the laboratory approaches in order to evaluate the response of the pore pressure buildup.

3. Nonlinear site response of liquefied areas

3.1 Introduction and studied area

Japan, Chile, the USA, Italy, Iran, and Turkey are some of the most important earthquake-prone countries, and they have been exposed to devastating activities over the last decades causing damaged buildings and many fatalities. One of the most essential regions for liquefaction studies in the literature is the Duzce/Sakarya Region in Turkey which was immensely affected by Kocaeli and Adapazari (Duzce) earthquakes in 1999. Liquefaction susceptibility in structural design should be considered in zones consisting of loose sand soils because it may cause significant damages as excessive settlement and sinking of the structures. This can only be possible by the nonlinear site-specific analysis.

In this section, the nonlinear site response of liquefied areas will be investigated using the well-documented in situ data taken from the city of Sakarya which is surrounded by Istanbul, Yalova, Duzce, and Bilecik. The region lays on the extension of the North Anatolian Fault Zone [25], and it is neighboring two local faults (Sapanca and Duzce), and the map is presented in **Figure 8**. The area suffered a lot, and many liquefied zone were inspected during Duzce Earthquake in 1999.

Five different boring log data were used in the analyses to evaluate the ground response in a liquefied prone zone. The soil profiles consist of almost 90% of silty sands (SM), and some low-plasticity clays, gravelly sands, and clayey sands can be found in very shallow depths. A summary table showing the information about the logs is presented below.

All the borings are 20 m long having shear wave velocity (V_s) values changing from 145 to 290 m/s. The average V_s of upper 30 m for each profile was also determined to be able to use it in the estimation of the surface response spectra

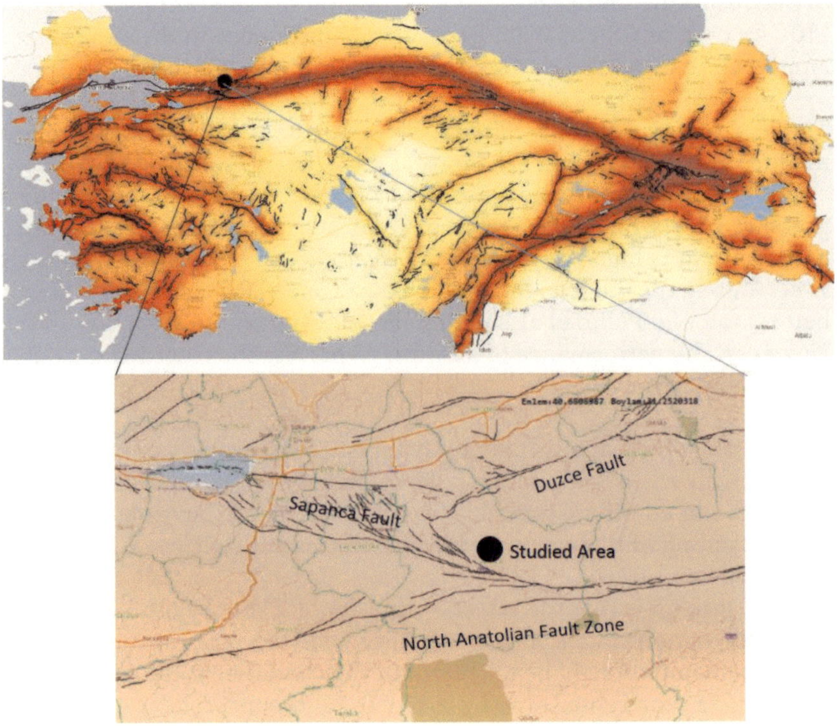

Figure 8.
Studied area and the active fault lines [29].

offered by the building codes, and the soil type of two borings is classified as ZE, and three of them are ZD. The fine content for the layers were obtained as well, and the variation is listed in the table. Layers have water content of 11–25%, and the unit weight of the soils samples laid in a range of 17.6–18.9 kN/m^3 throughout the profiles. Some triaxial tests were run in the laboratory to evaluate the strength of the sandy soils, and 33–38 degrees of friction angle along with 0–5 kPa cohesion were detected.

The dynamic properties (modulus reduction and damping behavior of soils at varying strains) of clays and sands were calculated adopting proposed models by Darendeli [26]. During the analyses, pore pressure generation was let to build up during cyclic loading in nonlinear analyses; however it was not possible by the frequency-dependent equivalent linear simulations. With the available information of the soil index properties, the required parameters were derived to model the pore pressure generation behavior suggested by Matasovic and Vucetic [27].

3.2 Motions used in the analyses

Building codes generally split possibility of earthquake levels into some groups, and the two of them have extensive use in geotechnical and earthquake engineering which are: (1) 2% chance of being exceeded in 50 years and (2) 10% chance of being exceeded in 50 years. They offer different response spectrum envelopes for each with the help of regional earthquake maps for the seismic design. To be used in the analysis, the possible local peak ground acceleration (PGA) was determined as 0.720 g for the studied area from the seismic maps which was just published in March 2018 by the Disaster and Emergency Management Authority (AFAD) under the Ministry of Interior in Turkey. Three different earthquakes happened in last

20 years (Duzce, Kocaeli, and Van earthquakes) were chosen to produce possible regional earthquake scenarios. The details of the earthquakes used in the process are shown in **Table 5**.

All of the earthquakes can be considered as the big magnitude earthquakes as seen in the table. Two of the events were taken from the Pacific Earthquake Engineering Research Center (PEER) [28] and one from the Strong Motion Data Base of Turkey (SMDB-TR) [29], and all of them were scaled to the local PGA accordingly. **Figure 9** shows the scaled versions of the earthquakes, in other words possible earthquake schemes for the studied area for the probability of 10% chance of being exceeded in 50 years.

The duration of the earthquakes differs from 25 to 85 seconds, and PGAs were set to be 0.72 g for all of them. More high-frequency content is seen from top to

Earthquake	Year	Source	Station	Magnitude	R_{jb} (km)
Duzce	1999	PEER	Duzce	7.1	0
Kocaeli	1999	PEER	Izmit	7.5	1.38
Van	2011	SMDB-TR	Van	6.7	19.2

Table 5.
Reference earthquakes used in the analysis.

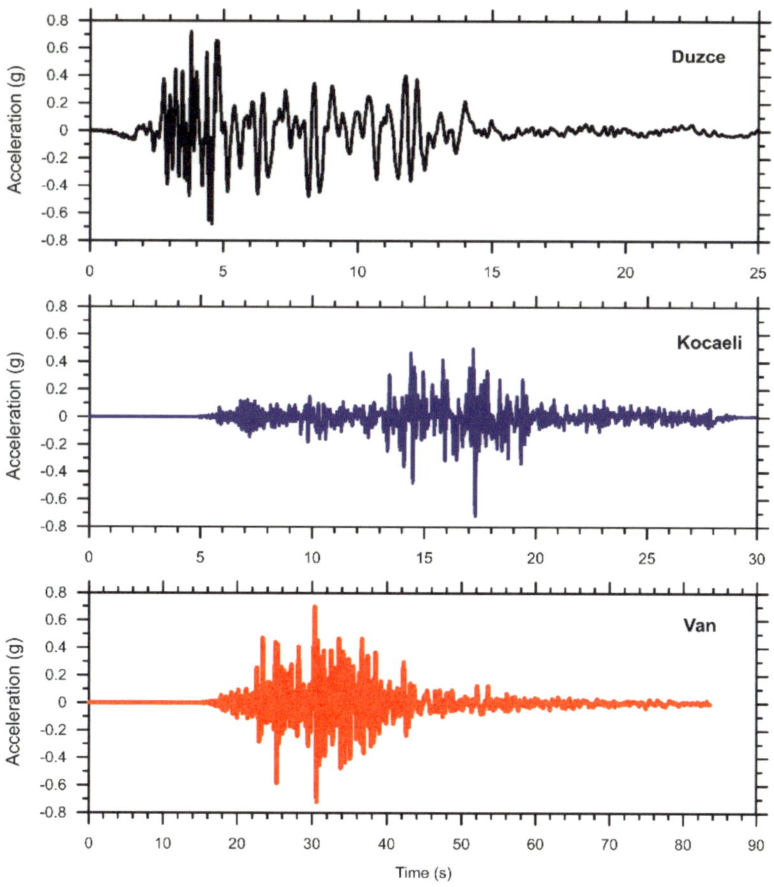

Figure 9.
Scaled versions of the earthquake data.

bottom of the figure, and these records were used as the base motions in the analyses.

3.3 Results of a specific soil profile

Site response analyses were conducted using DEEPSOIL [30] program version 6.1, and to be able to compare different approaches, five borings were excited at their base with three shakings with nonlinear and equivalent linear models for a total of 30 analyses. The results for the boring B1 shook by Duzce record, an example set, were presented in **Figures 10–12** showing the NL and EL model estimations for the surface acceleration, their corresponding spectral behavior on

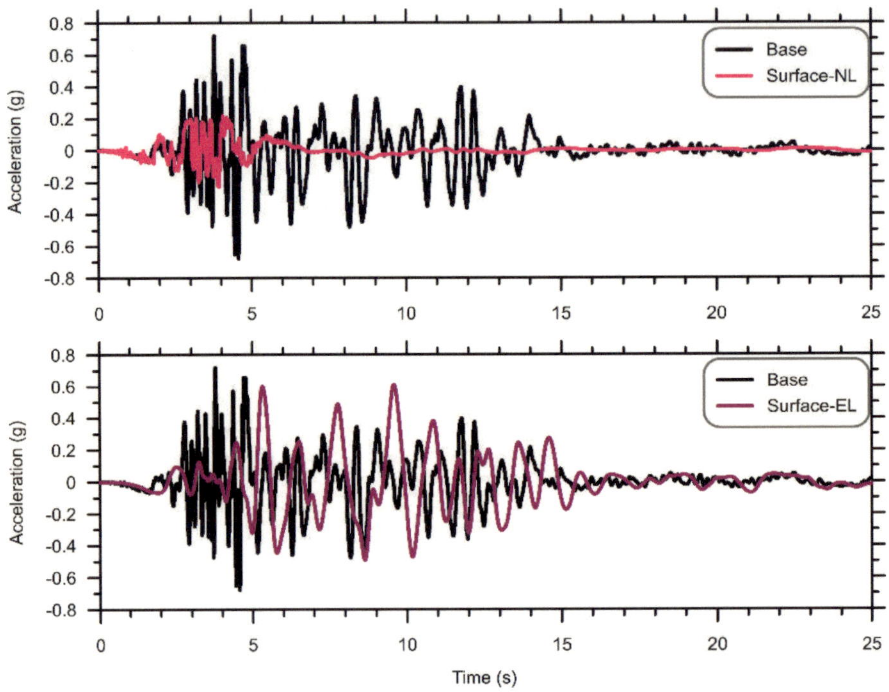

Figure 10.
Base and surface acceleration observed from the nonlinear (NL) and equivalent (EL) linear models.

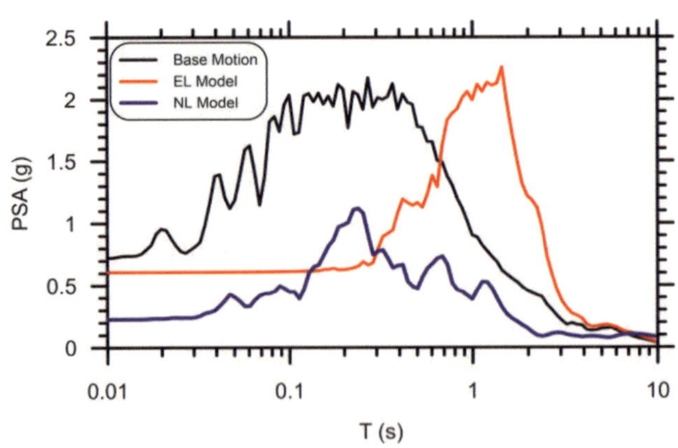

Figure 11.
Spectral acceleration of the base and the surface calculated from the NL and EL models.

the ground and the peak spectral accelerations, and maximum shear strains throughout the soil profile, respectively.

Regarding the surface acceleration predicted by two different approaches, there is a big deviation in the peaks and the time they occur. NL model shows that soil totally de-amplified the energy exerted by the motion, and the peak ground acceleration was noted around 0.25 g, whereas the peak-based acceleration was 0.72 g, and the main peaks happened around the similar time interval (3–5 seconds). However, an entirely different case was observed for the EL model. The prediction of the peak surface acceleration happened to be around 10 seconds extending the peak-based motion to 5–6 seconds, and not much de-amplification was noticed. In order to evaluate this behavioral diversity, the spectral acceleration at the surface was presented below.

The peak-based spectral acceleration (around 2 g) stays in a wide range of period between 0.1 and 0.8 seconds as seen in the figure. The NL model has compressed the behavior in terms of the size of peak ground and peak spectral accelerations. The model lessens the spectral acceleration 2–2.5 times at any period, whereas the EL model lengthened the peak spectral range to 0.9–1.8 seconds with almost no change in the amplitude. Another point about the EL model is that the initial flat part lasts longer than the base and NL model prediction. The last figure of the example set is the variation of the peak spectral acceleration and maximum shear strains through the soil column, and it is presented in **Figure 12**.

The difference between the models to estimate two parameters is concrete as observed from the figure. The peak-based acceleration was amplified a bit between 10 and 20 m and then de-amplified some for the rest of the profile by the EL model, whereas it was weakened all over the soil profile by the NL model. The soil profile was separated into three divisions, and it is supported by the related maximum shear strains at the same depths. It is obvious that two definite soil failures happened at the 8 m and 14 m according to the NL model, whereas the soil failures

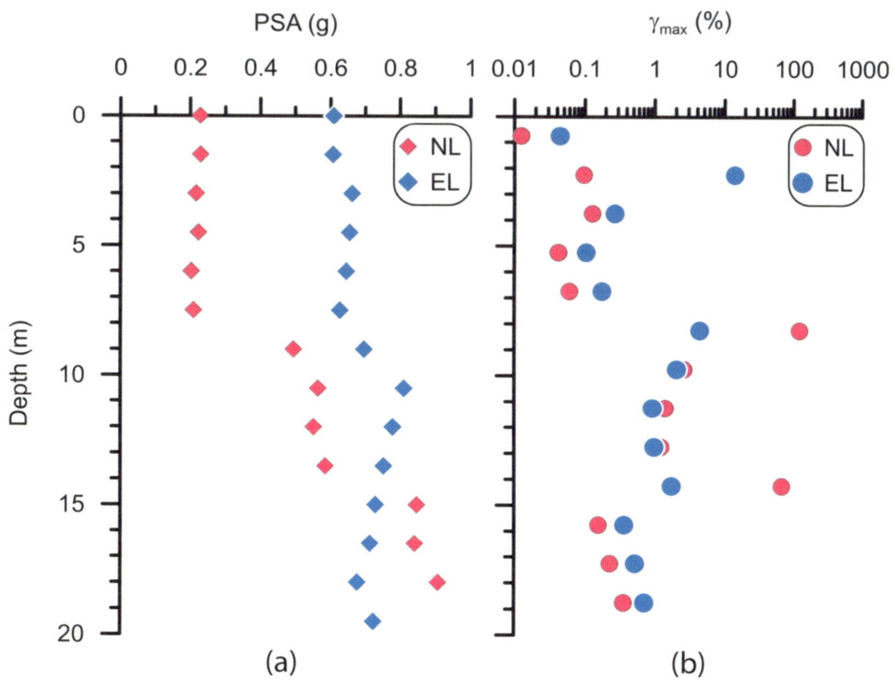

Figure 12.
Peak spectral accelerations and maximum shear strains throughout the soil columns for the NL and EL models.

seem to be predicted at the depth of 2 m and 8 m by the EL model. There are still other local failures between 9 and 13 m, but the deformation level is limited compared to the ones mentioned above. It can be concluded that not only the ground response predicted by two models is quite distinct, but also the seismic actions alter the seismic behavior extremely along the soil column.

3.4 Interpretation of the analyses

The loss of soil stiffness can be caused by either insufficient bearing capacity or liquefaction triggered by the excess pore pressure generated under cyclic loading. Thus, it is beneficial to see the pore pressure generation if it was the origin of excessive deformations during seismic excitations that only NL models are capable to investigate. Excess pore pressure ratio which is a value between 0 and 1 is one vital criterion in evaluation of the liquefaction. r_u of 1 means that the soil layer is under risk to liquefy for such type of loading. **Figure 13** shows the excess pore pressure ratios projected by the NL models for all the logs.

One clear remark can be pointed out here that the liquefaction occurring at various depths in different logs considering all three base motions sets out the possible explanation of the unreasonable deformation indicated in the previous figure. The NL model did not anticipate liquefaction triggering for the shallow depths which include some fine content, whereas there are at least nine points that soil layers mainly composed by silty sands are expected to lose their strength due to liquefaction for such sequence of events. The r_u value is greater than 0.8 for approximately 45 points meaning that even if there is no liquefaction initiated yet, soil would possibly lose its strength at these depths. Therefore, soil mitigation would be needed for possible construction in the area.

After liquefaction assessment of the soil profiles, the next step is to evaluate the site response in terms of nonlinear and equivalent linear approaches. Fifteen different analyses were conducted for each, and the peak spectral accelerations on the surface are presented in **Figures 14** and **15**. As mentioned before, building codes

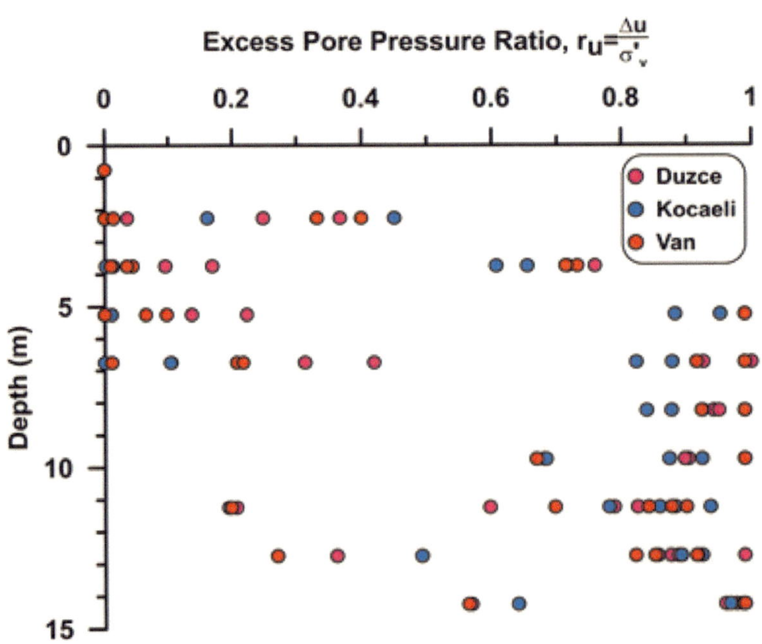

Figure 13.
Excess pore pressure ratios estimated from 15 NL analysis.

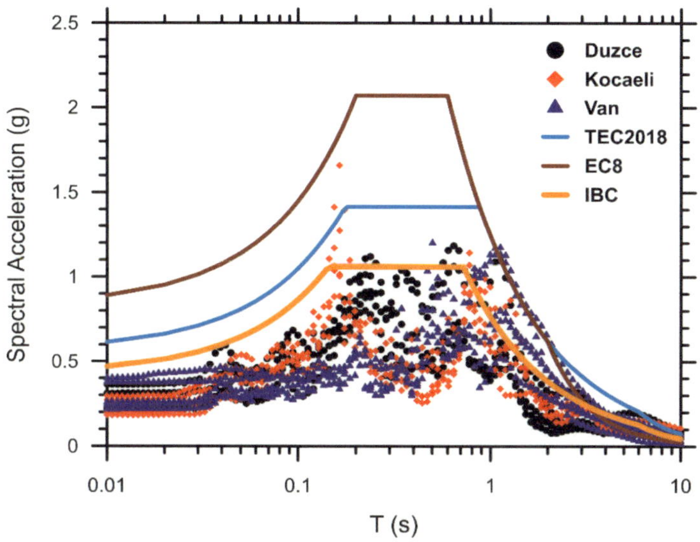

Figure 14.
Ground responses evaluated from the NL analysis and building codes' design spectra.

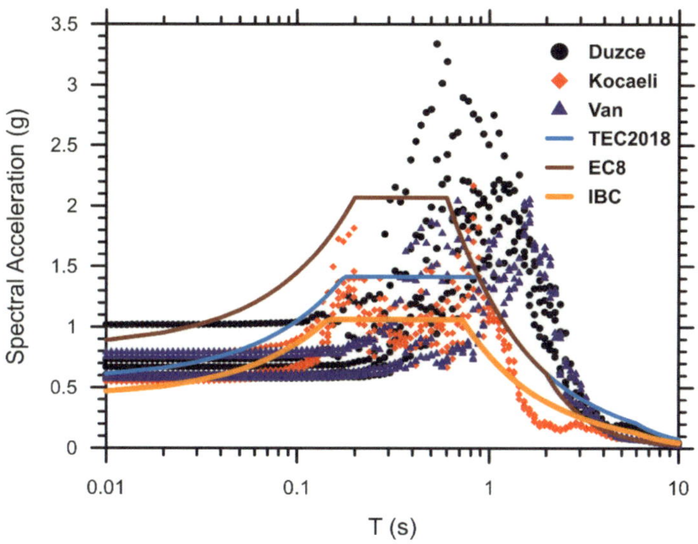

Figure 15.
Ground responses evaluated from the EL analysis and building codes' design spectra.

rule out the possible designs with seismic maps and some regulations. In order to see if they are good representatives of the site, corresponding response spectrum envelopes were evaluated using the new Turkish Building Code (TBC2018), Eurocode8 (EC8), and International Building Code-ASCE 7.05 (IBC), and they are shown in figures along with the ground response spectra gathered from the analyses.

Considering the PGAs, the variation is almost 100% from one log to another (0.20–0.40 g) in such a small area. The trend of the spectral accelerations is similar for increasing period; however the amplitudes differ a lot especially the change in peak spectral accelerations between 0.1 and 1 seconds. The lowest is reported as 0.3 g, whereas the highest is around 1.2 g with one outlier with the red dots. It is a bit off compared to the rest of the data; therefore it can be excluded to represent the general picture of the area.

In terms of the performance of the building codes with regard to expressing the site specifics, all of them are on the safer side (EC8 is the most conservative one and the IBC is the least) following the similar flow of spectral acceleration of the observed data from the nonlinear analyses. IBC's spectra would be considered as the most economical one compared to the others with missing some longer period content, and it would possibly be a better representation if the flat part of the spectra shifted on the right (by using larger corner periods). About the TEC2018, the spectra at higher period were estimated better, but the content from small to medium periods would not be cost-effective in the seismic design. As seen in the figure, EC8 envelope is not practical at all, and such design would be the most expensive one among three building codes.

Equivalent linear analyses are easy to perform with less parameter needed to constitute a site and widely used in the practice and literature. However, one should be cautious to use this approach in such cases. Since this frequency-based analytical approach is not suited for liquefaction analyses, the results may canalize the engineer to an inappropriate design. Although the details of the results obtained from the equivalent linear analyses will not be covered here, it is important to call attention to the differences. The spectral acceleration of 15 analyses is presented in **Figure 15** with their corresponding envelopes suggested by building codes.

As seen in the figure, the soil amplified the base motion and lengthened it to the higher periods for these earthquake scenarios. None of the building codes were capable of expressing the data very well, neither the peak spectral acceleration nor the corner frequencies. However, it should be underlined here that it is not the performance of the building codes, but instead it is the analytical approach of the equivalent linear analysis lacking to model the excess pore pressure generation and the damping behavior of the soil. Thus, nonlinear site-specific response in liquefied areas is crucial in the design, and the concept of liquefaction susceptibility should be explained more clearly in the building codes.

4. Conclusions

Pore pressure generation under cyclic loading is an important phenomenon for liquefaction triggering, and the number of cycles and/or duration to liquefy soil is affected a lot by the frequency content of the loading. Three different stress levels were tested at varying loading patterns, and it can be concluded that harmonic loading to estimate the liquefaction initiation is not that reliable in order to evaluate the dynamic behavior of sands. Regarding the number of cycles, it took almost 1848 cycles to liquefy the sand samples which is at least two times later than Type 5 loading, and this value goes up to 5.5 times at most for Type 6 at CSR = 0.15. The same comparison at CSR level of 0.20 shows that the shortest time to trigger liquefaction is 130 seconds, whereas it lasted 1110 seconds for harmonic loading being almost nine times later. Even among the irregular loading types, the difference was calculated as 2.5–3 times in terms of the number of cycles for soil to lose its stability at different stress levels. Therefore, frequency content and its link with the CSR level are of great importance on excess pore pressure buildup. With reference to the stress-based models in the literature to estimate the recorded data in the laboratory, the mentioned models were not that distinct to predict the data, and it was noted that the best prediction was noticed at the medium stress level (CSR = 0.2). Moreover, the model proposed by Baziar [16] had a better job presenting the test data than the other two.

Though nonlinear site response is a bit complicated to gather enough information to simulate the ground response, it is the most appropriate way to model the seismic behavior in liquefied zones. The in situ data was taken from Duzce area which was liquefied in 1999 in Adapazari earthquake, and the nonlinear analyses helped reproduce the liquefaction triggering at 45 points in different depths for varying earthquake scenarios. Since the equivalent linear approach cannot perform the liquefaction triggering, it alters the transmission of the shear waves affecting the spectral accelerations and the maximum shear strains throughout the soil layers. It was noted that the PSA was estimated as three times larger at shallow depths with corresponding shear strains as an example. Consequently, it should not be used for ground response in liquefaction-prone areas not to misinterpret the dynamic behavior of soils. Compared to three building codes to project the site response, the International Building Code is the most effective one to match the nonlinear analysis results.

Finally, it can be summarized that frequency content is highly effective on excess pore pressure buildup; the variation of its impact at different stress levels should not be disregarded in dynamic triaxial testing; and granting all this complexity of the nonlinear site-specific analyses, designers must be encouraged to run nonlinear analyses to model soil behavior better for safer superstructures, and the building regulations should be improved in terms of liquefaction susceptibility.

Author details

Kamil Bekir Afacan
Eskisehir Osmangazi University, Eskisehir, Turkey

*Address all correspondence to: kafacan@ogu.edu.tr

References

[1] Seed HB, Idriss IM. Analysis of soil liquefaction: Niigata earthquake. Journal of the Soil Mechanics and Foundations Division. 1967;**94**(SM3):83-108

[2] Seed HB. Landslides during earthquakes due to soil liquefaction. Journal of the Soil Mechanics and Foundations Division. 1968;**93**(SM5): 1053-1122

[3] Seed HB, Peacock WH. Test procedures for measuring soil liquefaction characteristics. Journal of the Soil Mechanics and Foundations Division. **97**:1099-1119

[4] Finn WDL, Lee KW, Martin GR. An effective stress model for liquefaction. Journal of the Geotechnical Engineering Division. 1977;**103**(GT6):517-533

[5] Finn WDL, Ledbetter RH, Wu G. Liquefaction in Silty Soils: Design and Analysis, Ground Failures under Seismic Conditions, Geotechnical Special Publication 44. New York: ASCE; 1994. pp. 51-76

[6] Castro G, Poulos SJ. Factors affecting liquefaction and cyclic mobility. Journal of the Geotechnical Engineering Division. 1977;**106**(GT6):501-506

[7] Castro G. On the behavior of soils during earthquakes—Liquefaction. In: Proceedings, NSF/EPRI Workshop on Dynamic Soil Properties and Site Characterization; EPRI NP-7337; Vol. 2; Palo Alto, California: Electric Power Research Institute; 1991. pp. 1-36

[8] Youd TL, Hoose SN. Liquefaction susceptibility and geologic setting. In: Proceedings, 6th World Conference on Earthquake Engineering; New Delhi; Vol. 3, 1977. pp. 2189-2194

[9] Youd TL. Geologic effects-liquefaction and associated ground failure. In: Proceedings, Geologic and Hydrologic Hazards Training Program. Open File Report 84-760, U.S. Geological Survey; Menlo Park, California; 1984. pp. 210-232

[10] Seed HB, Tokimatsu K, Harder LF, Chung RM. Influence of SPT procedures in soil liquefaction resistance evaluations. Journal of Geotechnical Engineering. 1985;**Ill**(12):1425-1445

[11] Kwan WS, Sideras S, El Mohtar CS, Kramer S. Pore pressure generation under different transient loading histories. In: Tenth U.S. National Conference on Earthquake Engineering Frontiers of Earthquake Engineering Anchorage; Alaska; 2014

[12] Lee KL, Albaisa A. Earthquake induced settlements in saturated sands. Journal of the Soil Mechanics and Foundations Division. 1974;**100**(GT4)

[13] Seed HB, Idriss IM, Makdisi F, Banerjee N. Representation of Irregular Stress Time Histories by Equivalent Uniform Stress Series in Liquefaction Analyses, EERC 75-29. Earthquake Engineering Research Center, University of California: Berkeley; 1975

[14] Booker JR, Rahman MS, Seed HB. A computer program for the analysis of pore pressure generation and dissipation during cyclic or earthquake loading. Berkeley: Earthquake Engineering Center, University of California; 1976

[15] Polito CP, Green RA, Lee J. Pore pressure generation models for sands and silty soils subjected to cyclic loading. Journal of Geotechnical and Geoenviromental Engineering. 2008; **134**(10):1490-1500

[16] Baziar MH, Shahnazari H, Sharafi H. A laboratory study on the pore pressure generation model for Firouzkooh silty sands using hollow

torsional test. International Journal of Civil Engineering. 2011;**9**(2):126-134

[17] Byrne PM. A model for predicting liquefaction induced displacement. In: Proceedings, 2nd International Conference on Recent Advances in Geotechnical Earthquake Engineering and Soil Dynamics; St. Louis, Missouri; Vol. 2. 1991. pp. 1027-1035

[18] Dobry R, Ladd RS. Discussion to soil liquefaction and cyclic mobility evaluation for level ground during earthquakes, by H.B. Seed and Liquefaction potential: Science versus practice, by R.B. Peck. Journal of the Geotechnical Engineering Division. 1980;**106**(GT6):720-724

[19] Dobry R, Ladd RS, Yokel FY, Chung RM, Powell D. Prediction of Pore Water Pressure Buildup and Liquefaction of Sands during Earthquakes by the Cyclic Strain Method, NBS Building Science Series 138. Gaithersburg, Maryland: National Bureau of Standards; 1982. 150 p

[20] Dobry R. Mohamad R. Dakoulas P. Gazetas G. Liquefaction evaluation of earth dams—A new approach. In: Proceedings, 8th World Conference on Earthquake Engineering; Vol. 3; 1984. pp. 333-340

[21] Vucetic M, Dobry R. Effect of soil plasticity on cyclic response. Journal of Geotechnical Engineering. 1991;**117**(1): 87-107

[22] Carlton B. An improved description of the seismic response of sites with high plasticity soils, organic clays, and deep soft soil deposits [thesis]. Berkeley: University of California; 2014

[23] Green RA, Mitchell JK, Polito CP. An energy-based pore pressure generation model for cohesionless soils. In: Proceedings: John Booker Memorial Symposium; 16–17 November 2000; Melbourne, Australia; 2000

[24] Kramer S, Hartvigsen AJ, Sideras SS, Ozener PT. Site response modeling in liquefiable soil deposit. In: 4th IASPEI/IAEE International Symposium. Effect of Surface Geology on Seismic Motion; 2011

[25] Komazawa M, Morikawa H, Nakamura K, Akamatsu J, Nishimura K, Sawada S, et al. Bedrock structure in Adapazari, Turkey—A possible cause of severe damage by the 1999 Kocaeli earthquake. Soil Dynamic And Earthquake Engineering. 2002; **22**(9–12):829-836

[26] Darendeli MB. Development of a new family of normalized modulus reduction and material damping curves [thesis]. Austin, Texas: Department of Civil, Architectural and Environmental Engineering, The University of Texas; 2001

[27] Matasovic N, Vucetic M. Cyclic characterization of Liquefiable Sands. ASCE Journal of Geotechnical and Geoenvironmental Engineering. 1993; **119**(11):1805-1822

[28] PEER. Pacific Earthquake Engineering Research Center. Available from: https://ngawest2.berkeley.edu/

[29] AFAD. Disaster and Emergency Management Presidency Earthquake Department. Available from: http://kyhdata.deprem.gov.tr/2K/kyhdata_v4.php

[30] Hashash YMA, Musgrove MI, Harmon JA, Groholski DR, Phillips CA, Park D. DEEPSOIL 6.1, User Manual. Urbana, IL: Board of Trustees of University of Illinois at Urbana-Champaign; 2016

Chapter 7

Weathered Granite Soils

Xirong Niu

Abstract

A great deal of weathered granite soils are distributed in mountainous areas around the world. With further improvements to the civil engineering, more and more construction infrastructures (roads, railways, dams, etc.) in mountainous areas will be built. Making full use of weathered granite soils, a type of special geomaterial, can alleviate the shortage of building materials in mountainous areas. Weathered granite soil has its own unique physical and mechanical properties, e.g., disintegrative, easy weathering, and particle breakage. In this chapter, a large number of field investigations and laboratory tests (including X-ray diffraction, sieving, heavy compaction, and large-scale triaxial) have been carried out. The process of weathering, the influence factors on particle breakage, and the mechanical properties of compacted weathered granite have been discussed. The results show that particle gradation, mineral content, blows per layer, and stress level have a significant effect on the particle breakage characteristics of weathered granite soils. The experimental results show that the product of the stress ratio at shear failure M_f and the stress ratio at characteristic state point M_c is not a constant but a power function of an average main stress p due to particle breakage. Hereby, the constitutive model of weathered granite soils was proposed.

Keywords: weathered granite soils, weathering mechanism, particle breakage, mechanical behavior, modified strength condition, constitutive model

1. Introduction

As a member of igneous rock family, granite is formed by magma inside the Earth. The main components of igneous rocks are quartz, feldspar, mica, and hornblende [1]. The features of granite include the full crystal-grain structure, hard texture and uniform material, and high compressive strength (120–200 MPa). Because joints exist in feldspar and mica, there are three groups of primary joints in granite. The significant difference (nearly doubled) between the expansion coefficients of quartz and feldspar causes granite surface to crack easily in the process of expansion and contraction. Hence, granite would weather easily, especially the one with coarse grain structure [2]. During the weathering process, granite is also influenced by long-term, geological tectonic movement, and climate changes in the specific geographical environment. Based on the different physical and chemical properties of granite mineral composition and the different degrees of weathering of granite, decomposed granitic soils are generally classified as weak weathering, weathering, strong weathering, and residual [3].

With further improvements to the civil infrastructure, more and more engineering construction projects (roads, railways, dams, etc.) will be built in complex mountainous areas. However, the shortage of building materials is an acute problem during constructing these infrastructures in mountainous areas. It is very important to study the engineering properties and applicability of special materials, e.g., weathered granite soil, in mountain area to solve the problem of shortage of building materials. Besides nonlinear stress-strain, elastic-plastic, dilatancy (shrinkage), and other properties, easy weathering and particle breakage are the distinctive and unique engineering property of weathered granite. The particle size distribution of the in situ soils is controlled by weathering process. Furthermore, the particle breakage characteristics of soils are affected by the particle size distribution. The particle size distribution of in situ weathered granite soil has been paid little attention in previous studies, but it is very important to have a full understanding of the particle size distribution of in situ weathered granite soil, because the particle size distribution has considerable influences upon engineering properties of weathered granite soil such as compatibilities, permeabilities, and strength-deformation characteristics [4–7].

In this chapter, the weathering mechanism, particle breakage, mechanical properties (including compaction characteristic, bearing characteristics, strength characteristics, and shearing-dilatancy characteristics), and constitutive model of weathered granite soils from Shanxi, China, were investigated. The results can provide a basis for the comprehensive understanding of the engineering characteristics of weathered granite soils and references for the utilization of weathered granite soils in engineering practice.

2. Specimen preparation and test methods

The weathered granite soil samples used in this chapter were obtained from a site near Mount Yunzhong (shown in **Figure 1**), a branch of the Lu-liang Mountains, on the eastern outskirts of Xinzhou City, Shanxi Province, China. The weathered granite soil was used as a part of the subgrade materials in the Xinbao highway (Xinzhou City to Baode City). There were three colors of the weathered granite in the field, red-brown, yellow-brown, and gray, as shown in **Figure 1**. The in situ dry unit weight and moisture content of the weathered granite were 17.82–18.22 kN/m^3 and 4.9–5.5%, respectively.

The excavation range of the filling materials used for Xinbao highway (Shanxi Province, China) was mined at a depth of less than 30 m on average under the

Figure 1.
Photograph of weathered granite samples.

surface of the Lu-liang Mountains. This is a general practice in producing subgrade materials for highways. Because the red-brown and the yellow-brown weathered granites were located at depths of less than 30 m, the lab tests in this chapter are focused on these two kinds of samples. The embankment is mainly filled with red-brown weathered granite, so parts of the tests in this chapter were mainly aimed at red-brown samples.

Because the physical meaning of Fukumoto's weathering model [4] is clear and easy to use, this model was used in this chapter to evaluate the weathering process of granite. After the thorough investigation on the distribution of in situ weathered granite, numerous sieving tests on a large number of granite samples obtained from typical sections A and B (shown in **Figure 2**) were carried out. According to Fukumoto's grading model, geological year's parameter m and geometric progression constants r of granite samples at different depths were calculated. The experimental tests involved in this chapter, including X-ray diffraction, sieving, heavy compaction [8], and large-scale triaxial test [9], were investigated to research the particle breakage characteristics of compacted weathered granite.

Figure 3 shows the composition of the rock samples in this study. **Figure 3** shows that the colors of the samples became gradually brighter with an increase in the quartz content. The increase in potassium feldspar or feldspar content was the

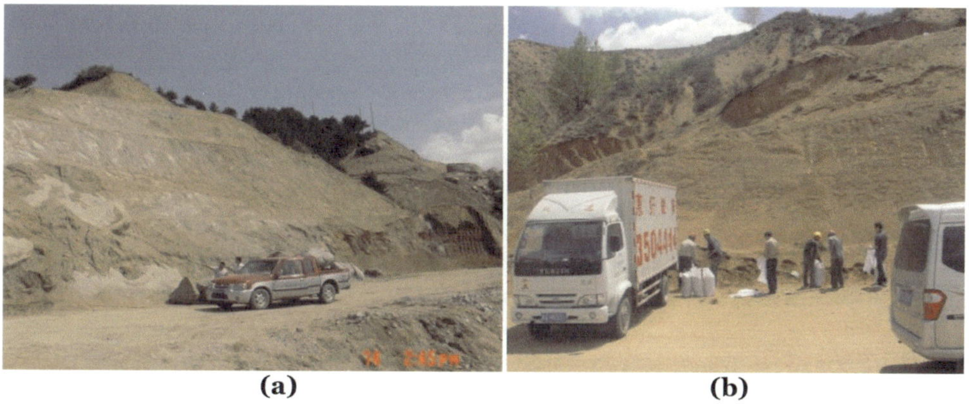

Figure 2.
Photograph of field sampling: (a) section A and (b) section B.

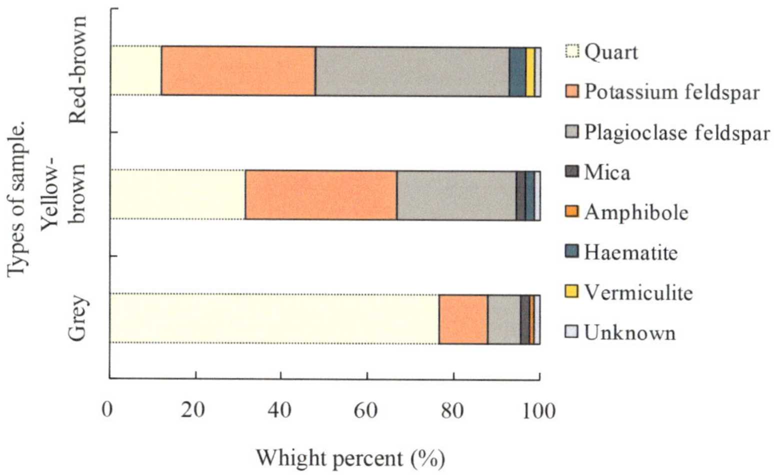

Figure 3.
Mineral composition of weathered granite samples.

main cause of the brighter colors of the rocks. Geng et al. [10] noted that the amphibole was bluish-green, and the biotite was yellow in Mount Yunzhong. In addition, Geng et al. [10] described the main body of the granite as a coarse grain (porphyritic granite), with fine-grained granite on the edge. The three types of samples used in this chapter had quartz (mean 40.3%) and feldspar (mean 54.3%) as the dominant minerals present with smaller amounts of mica (mean 1.3%), amphibole (mean 2%), hematite (0.5%), vermiculite (1%), and unknown (1%) materials.

3. Weathering process

Based on the geometric fractal theories and some assumptions, Fukumoto [4] proposed a gradation equation for decomposed granite soils by using mathematical statistics method and made it possible to quantificationally describe the weathering process of granite by using a certain mathematical model to evaluate the particle breakage of weathered rock. Because Fukumoto's model is more scientific and easier to use, this model has been widely used in describing the process of rock weathering. Thus, this model was also used in this chapter to describe the weathering process of in situ weathered granite.

Figure 4 shows changes of parameters m and r for the samples at different depths of mountain profile, h.

The results of statistical analysis indicate an approximate power function relationship between h and m (or r), and the correlation of relationship was relatively good. The power function relationships between h and m (or r) in section A are shown as follows:

$$m_A = 3.324h^{-0.706} \tag{1}$$

$$r_A = 0.243h^{0.153} \tag{2}$$

where h is the depth of granite samples, which was measured from the surface of mount, m_A is the geological year's parameter in section A, and r_A is the geometric progression constant in section A.

The power function relationships between h and m (or r) in section B are shown as follows:

$$m_B = 3.729h^{-0.747} \tag{3}$$

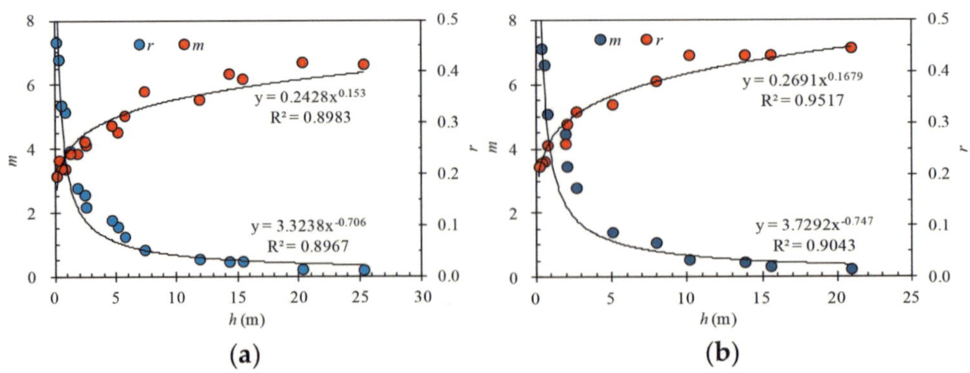

Figure 4.
Change of parameters m and r for samples at different depths: (a) section A and (b) section B.

$$r_B = 0.269h^{0.168} \quad (4)$$

where m_B is the geological year's parameter in section B and r_B is the geometric progression constant in section B.

Actually, the geological year's parameter, m, reflects the degree of geological actions (including physical actions and chemical actions) [11]. The deeper the granite is buried, the less the geological action will affect it. The above results indicate that m was close to zero at a depth of 20–25 m under the surface of the mount in sections A and B, which indicate that the physical and chemical actions on the granite in this research can be negligible at this depth. The geometric progression constant, r, represents the integrality of rock. The r of section A and section B studied in this chapter was close to a certain value at a depth of 25 and 20 m, respectively, which indicates that the granite in Mount Yunzhong was close to the complete unweathered state at a depth of 20–25 m.

It can be found from the results that the m of section A was larger than that of section B at the same depth, indicating that the ability of section A to resist geological and climatic environment was worse than that of section B. On the other hand, the r of section A was smaller than that of section B at the same depth, which illustrates that the integrality of granite in section A was worse than that in section B [11]. Although section A and section B are under the same geological and climatic environment, the values of m (or r) are different between these two sections. The main reason for the difference was the difference in internal properties (e.g., the mineral composition of rocks) of weathered granite between these two sections. Of course, some external factors could not be excluded; for example, the probability of fracture distribution of section A is larger than that of section B.

4. Particle breakage characteristics

Many aspects, such as particle gradation, mineral composition, blows per layer, and stress level, can influence the characteristics of particle breakage of compacted weathered granite.

Hardin [12] proposed the concept of relative breakage (B_r), which was achieved from the variation in particle size distribution curves before and after loading. Relative breakage can embody the overall change in particle size distribution before and after loading and overcomes the shortcoming of using single index of sieving test results (percentage of particles passing a given sieve size or sieve size corresponding to a given percentage passing) as the evaluation index for particle breakage [13–16]. In view of the above, B_r has been widely used in evaluating soil particle breakage characteristics, because it can represent the basic attribute of the soil material and is more scientific and reasonable to use.

4.1 Particle gradation

Standard heavy compaction tests on four red-brown samples and four gray samples of weathered granite with different gradations under optimum moisture content conditions were investigated. Sieving tests were implemented on the samples before and after compaction, and the uniformity coefficient ($C_u = d_{60}/d_{10}$) and relative breakage (B_r) of the samples before and after the compaction test were calculated. The relationship between C_u and B_r is shown in **Figure 5** [11]. The test results show that B_r decreased with the increase of C_u which indicates that as the C_u of weathered granite increased, the degree of particle breakage was reduced because of the enhancement of the intergranular locking effect. In addition, it can be seen

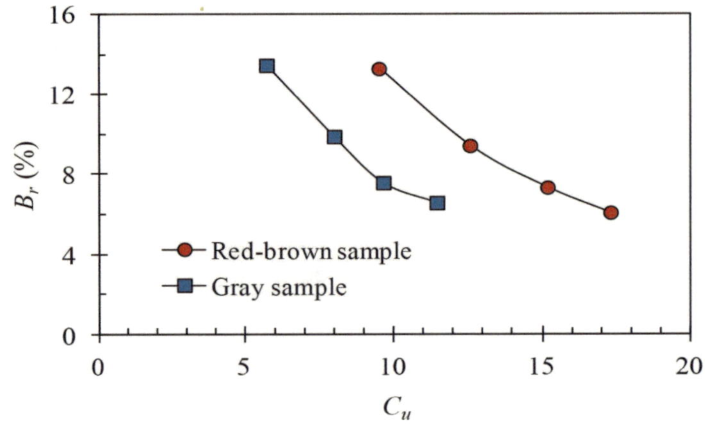

Figure 5.
B_r-C_u curves.

that the rate of decrease of relative breakage of the red-brown sample was higher than that of the gray sample, reflecting that because the degree of weathering of red-brown weathered granite is higher than that of gray weathered granite, the strength of the red-brown sample was less than that of the gray sample. This shows that the sensitivity of the red-brown sample to particle heterogeneity was greater than that of the gray sample [11].

4.2 Mineral content

The red-brown, yellow-brown, and gray weathered granite samples were prepared with approximately the same particle size distribution. Standard heavy compaction tests were performed on the samples under optimum moisture content conditions [11]. The results of sieving before and after the test are shown in **Figure 6**.

The relationship between relative breakage and quartz (or feldspar) content was analyzed by calculating the relative breakage of different samples before and after

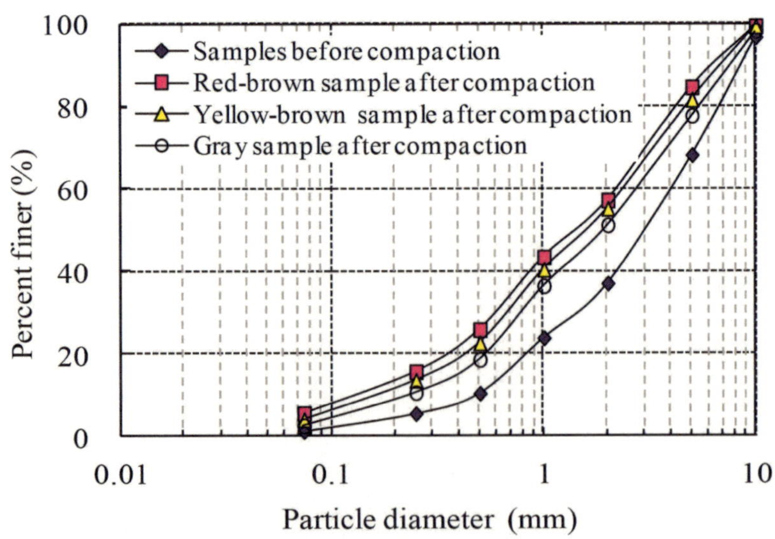

Figure 6.
Sieving curves of different samples before and after the test.

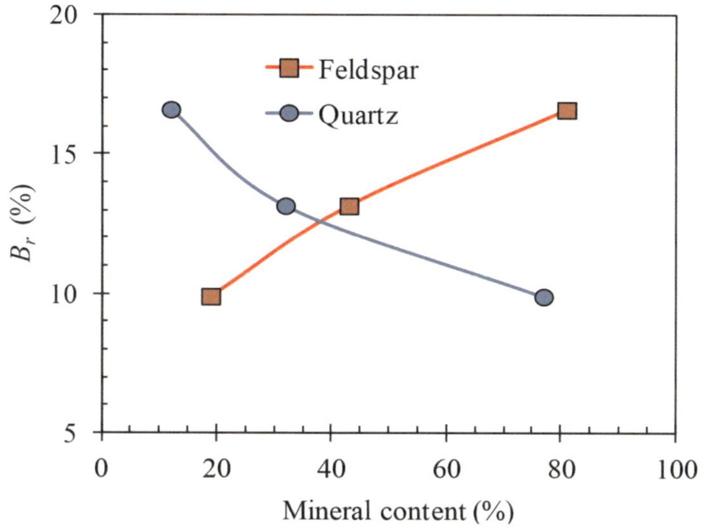

Figure 7.
Relationship curve between mineral content and B_r.

the standard compaction test. **Figure 7** shows the relationship between relative breakage and quartz (or feldspar) content. It can be seen from the figures that B_r decreased with the increase of quartz content, while B_r increased with the increase of feldspar (=plagioclase feldspar + potassium feldspar). The results show that quartz and feldspar content have an obvious effect on the particle breakage characteristics of weathered granite [11].

The main reasons that the B_r of samples with high quartz content after compaction test was small are as follows: (1) the probability of breakage of samples with more quartz content is small because quartz has a high strength, (2) samples with high quartz content have strong ability to resist being weathered, and (3) there are few microcracks in samples with high quartz content. On the contrary, feldspar has little strength and is easily weathered, so the samples with higher feldspar content showed obvious particle breakage characteristics [11].

4.3 Blows per layer (compaction degree)

In order to analyze the effect of blows per layer of samples on the particle breakage properties of weathered granite, four red-brown weathered granite samples with the same initial particle size distribution were prepared, and four different heavy compaction tests were conducted, with blows per layer (BPL) of 30, 50, 75, and 98 [11]. The sieving results before and after the tests are shown in **Figure 8**.

As shown in **Figure 9**, the relative breakage increased with an increase in blows per layer, but the increasing level of relative breakage decreased. Furthermore, with the further increase of blows, B_r tended toward a certain limit value. Considering the engineering practice, the compaction degree of samples was analyzed, and the compaction degrees of samples and corresponding relative breakage are depicted in **Figure 10**. It can be concluded from this figure that there is an approximate linear growth relationship between compaction degree and B_r. It can be found that for the same weathered granite fillings, relative breakage can be used to reflect the compaction performance indirectly on the basis of the relationship between compaction effect and compaction performance. Excessive compaction may lead to excessive particle breakage of soils and is not conducive to the long-term stability of subgrade [11].

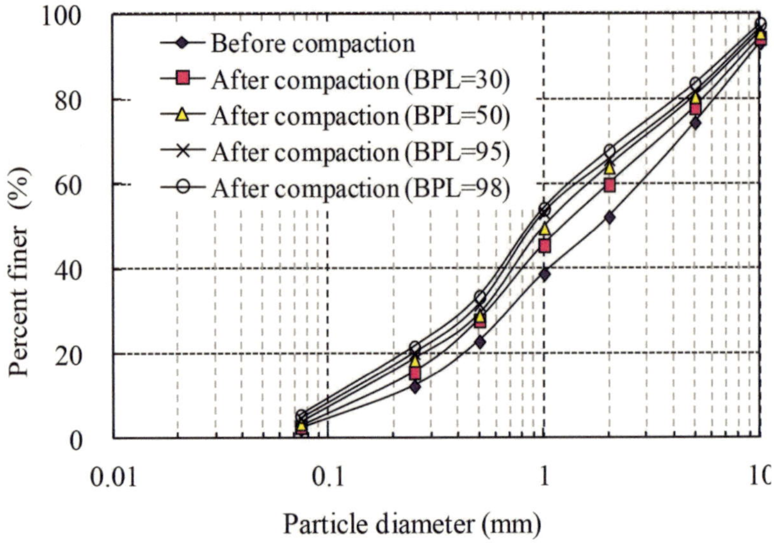

Figure 8.
Sieving curves of samples under different blows per layer. BPL: blows per layer.

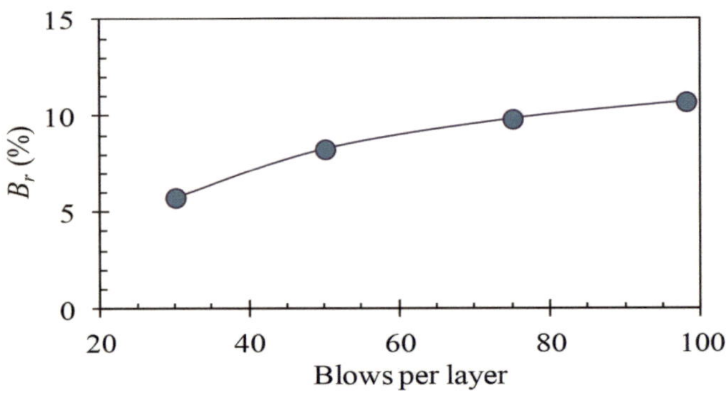

Figure 9.
Relationship curve between blows per layer and B_r.

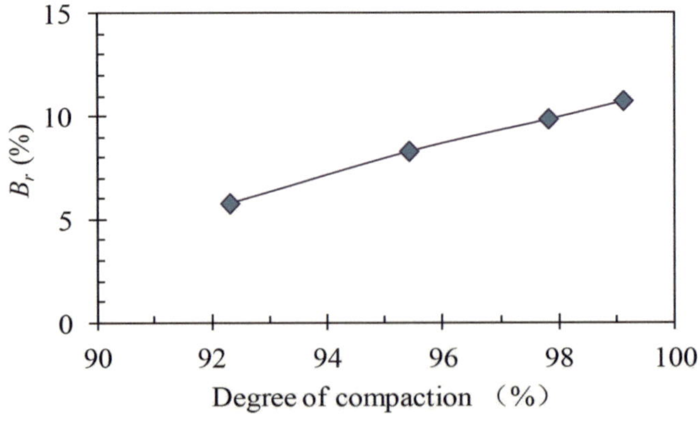

Figure 10.
Relationship curve between degree of compaction and B_r.

4.4 Stress level

The breakage of soil particles in a sample of rockfill under moderate stress will be quite evident [11]. For the purpose of studying the particle breakage characteristics of compacted weathered granite soil under different stress levels, particle size analysis tests were conducted on the samples before and after consolidated drained large-scale triaxial tests, and the relationship between relative breakage and confining pressure (or stress ratio, q/p) was analyzed [11].

Because specimens for the triaxial test must be prepared by the vibrating forming method, particles of the specimens may be partially crushed. In order to improve the accuracy of the experiment, two specimens for the large-scale triaxial test under the same confining pressure were artificially prepared with approximately the same particle size distribution before vibrating compaction. One of the samples was sieved after vibrating compaction and before the triaxial test. The results of the sieving test for this sample were used as the initial gradation of the other sample for the triaxial test [11].

From the stress-strain curve of samples in this study, it can be seen that when the axial strain reached 15%, the residual strength of the specimen under different confining pressures reached its constant value. Therefore, when the axial strain reached 15%, the triaxial tests were forced to stop, and then the sieving tests were performed. **Figures 11** and **12**, respectively, show variations in relative breakage (B_r) with confining pressure (σ_3) and stress ratio (q/p) [11].

As indicated in **Figure 11**, at the end of the triaxial test, the relative breakage of samples increased with the increase of confining pressure, while the increasing amplitude of relative breakage decreased slightly with the increase of confining

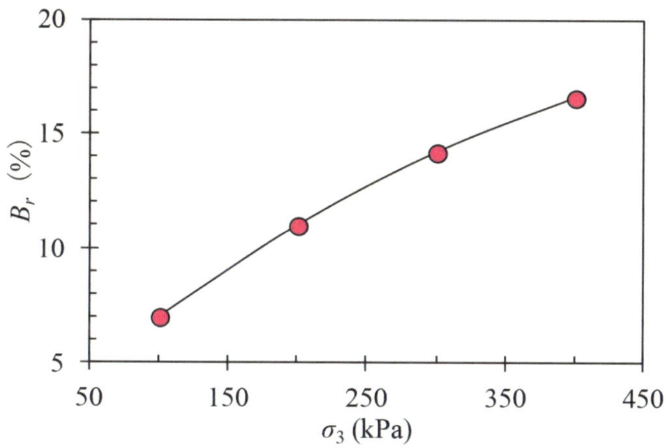

Figure 11.
B_r-σ_3 curves.

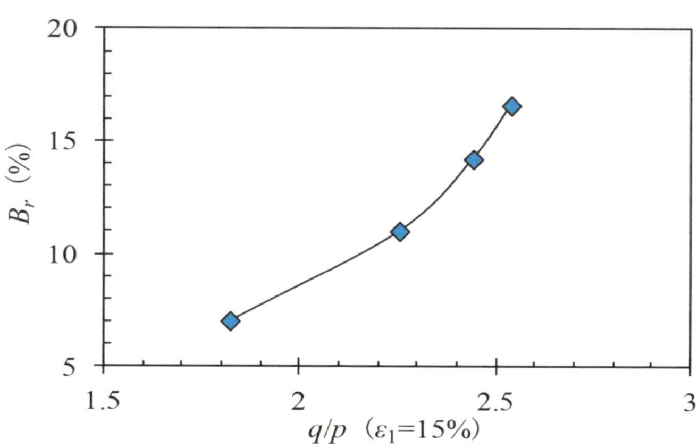

Figure 12.
B_r-q/p (ε_1 = 15%) curves.

pressure. It can be considered that the higher confining pressure limited the further compaction and movement of soil particles in soil samples and resulted in producing microcracks and fracture within a single particle. The data in **Figure 12** illustrate that at the end of the experiment (ε_1 = 15%), both relative breakage B_r and the amplitude of B_r increased with the increase of the stress ratio q/p. These results indicate that under the same axial strain conditions, more particles were broken due to the increased strain confinement caused by higher confining pressure [11].

5. Mechanical behavior

5.1 Compaction characteristics

Figure 13 shows the compaction curves for the red-brown and yellow-brown samples with different clay contents. It can be seen from **Figure 13** that the peak maximum dry density (MDD) values occur most significantly when the clay contents are in the range of 7.5–10%. The peak MDD of the red-brown and the yellow-brown samples of weathered granite soil is 2.32 and 2.38 g/cm^3, respectively. The results of the tests clearly show that as the clay content increases, the MDD tends to be considerably reduced after it reaches the peak MDD. Therefore, the experimental results show that the clay content of weathered granite soil has a remarkable influence on its compaction characteristics. Furthermore, at peak MDDs, the clay content of red-brown weathered granite soil is 1% larger than that of yellow-brown weathered granite soil. Because the particles of the red-brown samples were smaller than the ones in the yellow-brown samples, the yellow-brown samples mixed with clay could be easily formed into the suspended-dense structures in the process of compaction. The red-brown samples mixed with clay could be easily formed into the skeleton-dense structures. Consequently, the peak MDD of red-brown samples was smaller than that of yellow-brown samples. In addition, the higher the clay content, the larger is the optimum moisture content, and this relationship is approximately linear. These results also indicate that because the gradation of red-brown samples is finer than yellow-brown, the optimum moisture content (OMC) of red-brown samples is 0.1% smaller than yellow-brown. It is noted that the

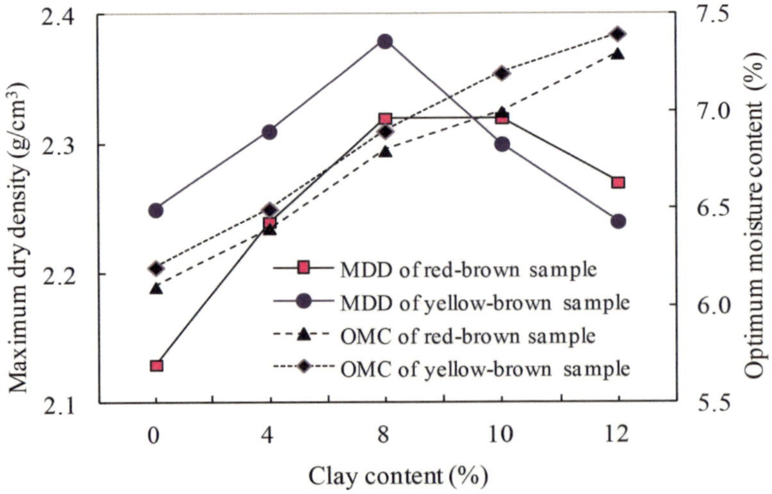

Figure 13.
Compaction curves for weathered granite samples with differing clay content.

increase of OMC slowed down slightly when the clay contents of the samples were more than 8% [17].

5.2 Bearing characteristics

Figure 14 represents the relationship of clay contents and CBR (California Bearing Ratio) obtained from different compaction tests, with the blows per layer of 30, 50, and 98, respectively. As shown in **Figure 14**, the CBR comes to a peak value when the value of blows per layer is 50 or 98. Because the compaction power is not enough to make rock material in a dense state when compacted under 30 blows, the CBRs of the samples with 30 blows do not arise to a peak value. However, with the increase of clay content ratio, the samples with 30 blows will eventually arise to a peak CBR. The experimental results show that the peak value of CBR increases with the increase of blows per layer. But the results also show that as the blows per layer increase, the clay content ratio at the point of peak CBR decreases. It is noted that the clay content ratio at the point of peak CBR with 98 blows per layer is approximately 4%, which is 4% less than the clay content ratio at the point of peak maximum dry density. The cause of the above results is mainly because that when clay content ratio exceeds a certain value (i.e., 4%), the interlocking structure of the compacted weathered-granite would be opened by the clay in the material, the internal friction angle (φ) would be decreased, and the penetration resistance subsequently would be declined. Finally, the clay content ratio at the point of peak CBR is larger than that the ratio at the point of peak maximum dry density [17].

5.3 Strength characteristics

Figure 15 shows the typical stress-strain relationship of the CD tests on four representative red-brown samples of pure weathered soil under four different confining pressures, where σ_1 is the axial stress, σ_3 is the confining pressure, and $d\varepsilon_1$ is the axial strain. **Figure 16** shows this typical ε_v-ε_1 relationship of the CD tests, where ε_v is the volumetric strain.

It can be found in **Figure 15** that the peak deviator stress increases as the confining pressure increases. The internal particles of weathered granite soil could be overturned, stridden, and dislocated under lower confining pressure and could

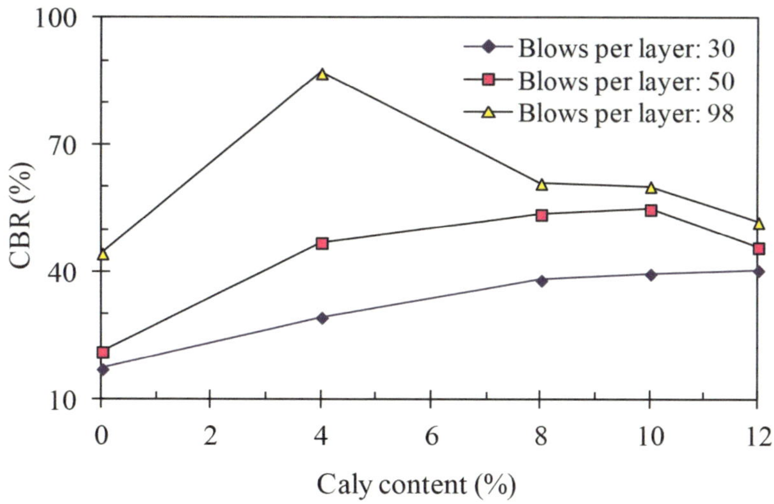

Figure 14.
Change in CBR for soils with differing clay content.

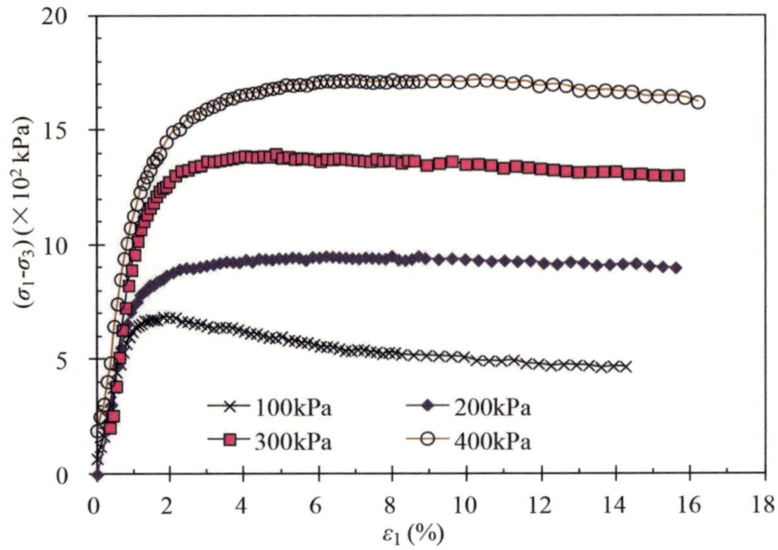

Figure 15.
Stress-strain curves.

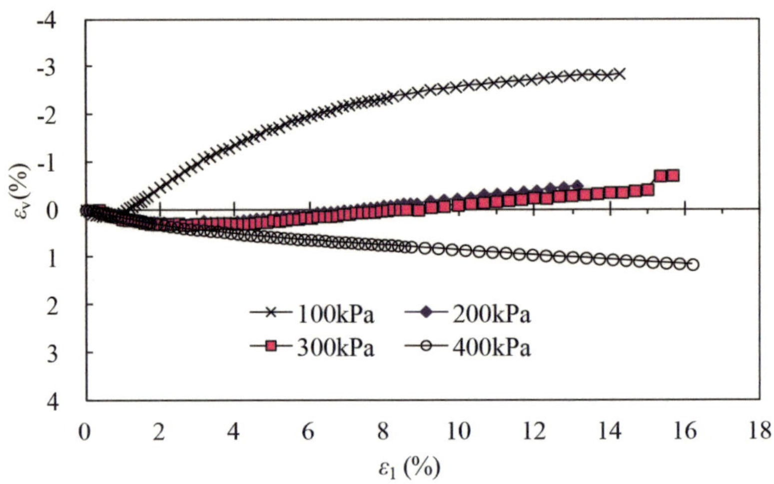

Figure 16.
Curves of ε_v-ε_1.

be crushed, filled, and compacted under higher confining pressure. It was eventually manifested that the specimens transit gradually from strain softening status (e.g., when the initial confining pressure was 100 kPa) to strain hardening status (e.g., when the confining pressure was 200 kPa) with the increase of confining pressures on the samples. At any event, there exists a critical confining pressure. The samples soften in strain when the experiment confining pressures were lower than the critical confining pressures and harden in strain when the experimental confining pressures were higher than the critical confining pressures [18]. Thereafter, it could be deduced that the critical confining pressure in this study was between 100 and 200 kPa. It can be observed from **Figure 16** that the samples transit generally from shear dilation to shear shrinkage, and the volumetric strain of the transition point of the specimens increases gradually with the increase of confining pressures on the samples. In summary, as the confining pressure increases, the volumetric strain at failure changes toward volume contraction. According to the results of large-scale triaxial tests, the Mohr's circles and the strength failure envelope can be depicted in **Figure 17** on the basis of the Mohr-Coulomb strength criterion. The linear Mohr's envelope of the weathered granite soil in this work can be expressed as

$$\tau_f = \sigma \tan 41.3° + 0.55 \text{ MPa}, \tag{5}$$

where τ_f is the peak shear stress and σ is the normal stress.

Eq. (5) shows that the cohesive strength of the samples is 55 kPa. However, because the weathered granite soil belongs to coarse grained soil, the cohesive strengths of the samples should be close to zero. It has been proved that the particles of coarse grained soil might be broken and its peak strength is nonlinear [19]. **Figure 18** shows that the ratio (τ_f/σ) is not constant but decreases when σ increases. In addition, the Mohr envelope is bending downward. The radius of bend reduces as the confining pressure increases, which reflects that the particle breakage is controlled by the confining pressures noticeably.

According to De-Mello [20], the power function of weathered granite soil in this study can be expressed as follows:

$$\tau_f = 0.361\sigma^{0.803} \tag{6}$$

As shown in **Figure 18**, the fitting envelope has a good correlation, which further explains that weathered granite soil has the property of nonlinear strength because of particle breakage. The nonlinear strength characteristics of compacted weathered granite soils could be affected by rock strength, intergranular friction, occlusal effect, etc. The mechanism of the nonlinear strength could be interpreted by means of the variation of failure surface angle in previous literature [21]. Under lower confining pressure, the internal friction angle was relatively large, the bending of the strength envelope was obvious, and the samples had the obvious characteristics of shear dilation and particle breakage. With the increase of confining pressure, the internal friction angle decreased and the bending degree of strength envelope decreased gradually. It was showed that under higher confining pressure, the strength characteristics of compacted weathered granite soils were mainly dominated by particle breakage.

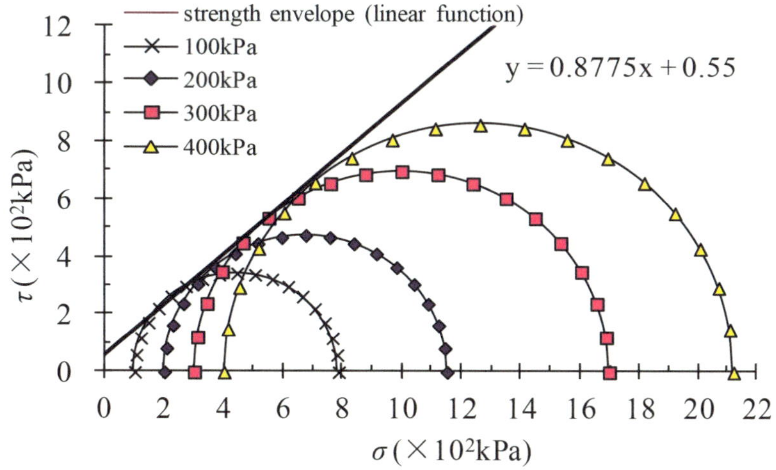

Figure 17.
Mohr strength envelope of red-brown samples.

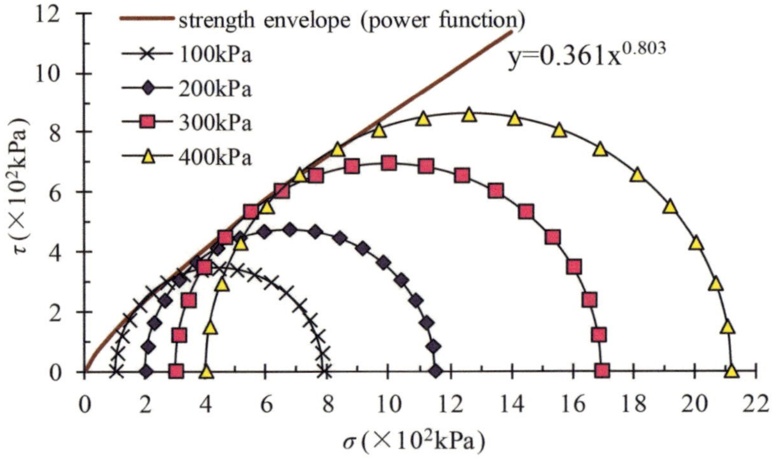

Figure 18.
Actual strength envelope of red-brown samples.

5.4 Shearing-dilatancy characteristics

Weathered granite soil, as coarse grained soil, has obvious shear dilation property, which was found through the CD tests in this chapter. **Figure 19** shows the results of triaxial tests under difference confining pressures with respect to the relation between the stress ratio of q/p and the strain increment ratio of $d\varepsilon_v/d\varepsilon_1$. The tests were performed according to the stress and strain increment parameters used in widely accepted models, such as Cam-clay model, where $q = \sigma_1 - \sigma_3$ and $p = (\sigma_1 + 2\sigma_3)/3$. **Figure 19** shows the following observations: (1) under difference confining pressures, the strain increment ratio is always larger than zero at the initial point of shear. The ratio decreases with the increase of stress ratio q/p, which indicates that the volumetric strain of weathered granite soil changes toward the volume contraction. At the same time, the contraction ratio in the initial stage of triaxial shear is larger than that in the following stage. The reason for this

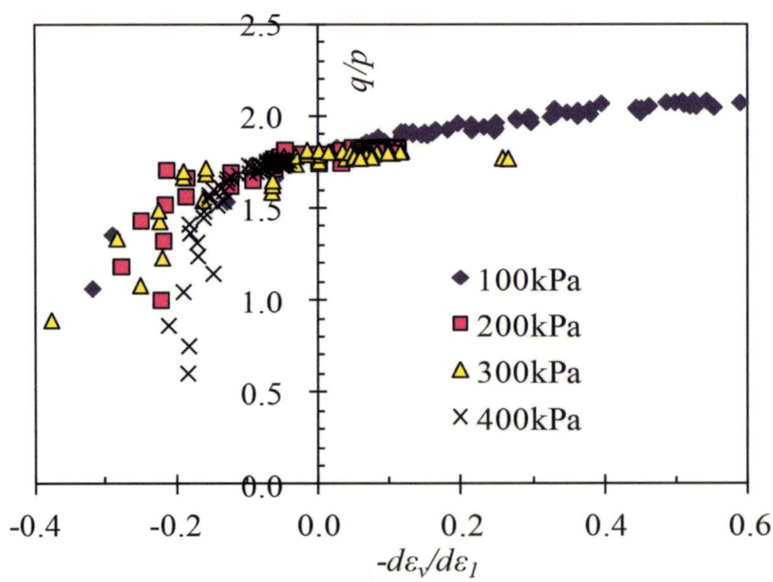

Figure 19.
q/p: $d\varepsilon_v/d\varepsilon_1$ of sample data with different confining pressures.

phenomenon is that once the soil reaches to a certain compaction rate, it gets difficult to compact further. (2) When the confining pressures change from 100 to 300 kPa, the strain increment ratio decreases as shear stress increases. The samples reached the stage transition point ($d\varepsilon_v = 0$) when the strain increment ratio was equal to zero at certain level of stress. After the stage transition point, with the increase of shear force, the $d\varepsilon_v/d\varepsilon_1$ ratio becomes negative. At this time, the volumetric strain of weathered granite soil changes toward volumetric contraction dilatancy. The aforementioned phenomena reflect the characteristics of dilatancy under lower confining pressures. The cause of the phenomena is due to the situation that the internal particles of weathered granite soil might be overturned and dislocated. (3) When the confining pressure is equal to 400 kPa, the $d\varepsilon_v/d\varepsilon_1$ ratio is always larger than zero during shearing processes. It indicates that interparticle contact stresses in the weathered granite soil increase gradually and the particle breakage becomes obvious under larger confining pressures. At this stage, the samples were made further denser as the broken particles filled into the voids of soil and the shrinkage behavior of soil samples is obvious.

Based on the above test results, there must exist a confining pressure between 300 and 400 kPa that makes the volumetric strain increment to be equal to zero. In other words, between the aforementioned observations (2) and (3), there must exist a relatively balanced state when the dilatancy caused by the overturn and dislocation of soil particles is equal to the shrinkage caused by the breakage and compaction of soil particles.

The curves of σ_1/σ_3 versus ε_v in **Figure 20** show the relationships between σ_1/σ_3 and ε_v. The relationships include two cases. One case is that the curve bends to left. The other case is that the curve bends to right. The volumetric strain of samples changes to the volume contraction initially and dilatancy later with the increase of the stress ratio of q/p under lower confining pressures, as shown in the former case when the curve bends to left. The volumetric strain of samples changes always toward the volume dilatancy with the increase of the stress ratio of q/p under larger confining pressures in the latter case when the curve bends to right.

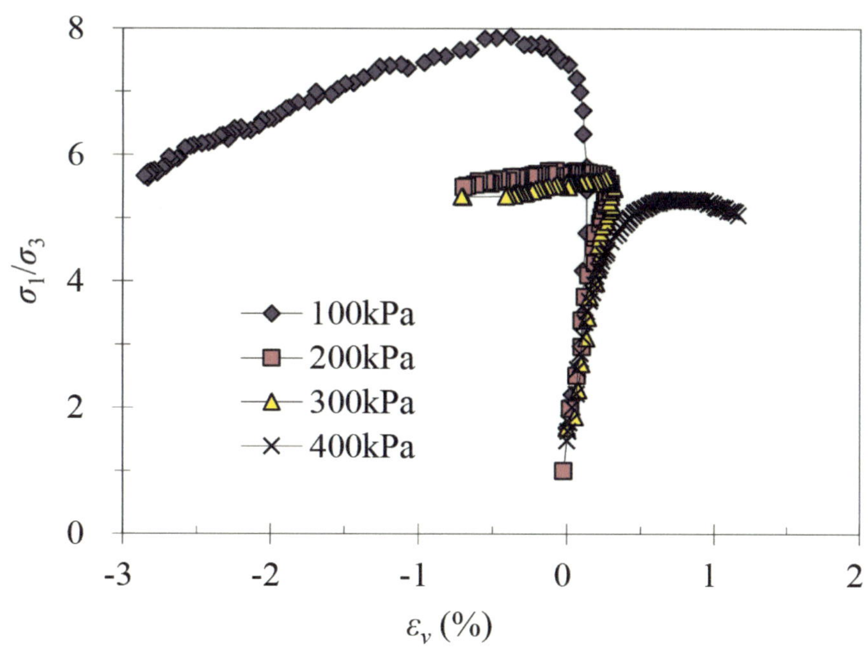

Figure 20.
Curves of σ_1/σ_3 versus ε_v.

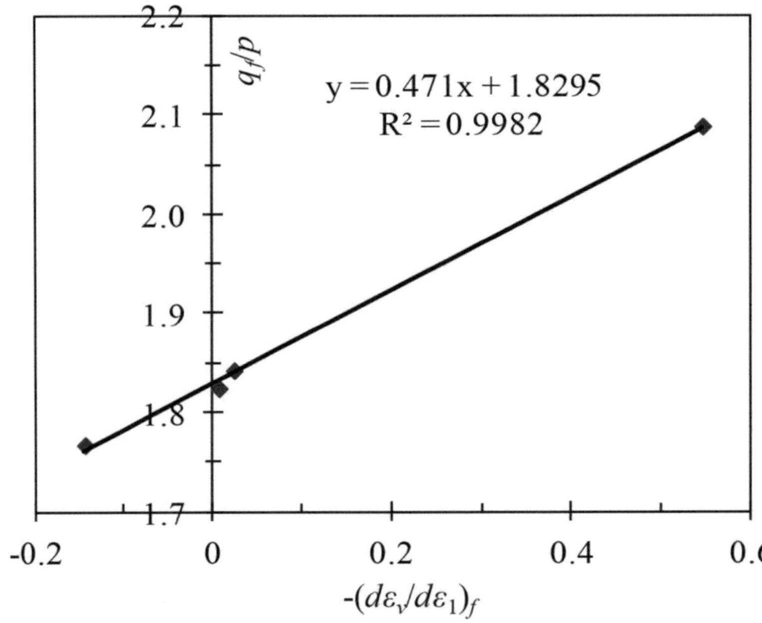

Figure 21.
Curve of q_f/p versus $-(d\varepsilon_v/d\varepsilon_1)_f$.

Figure 21 shows the relationship between peak stress ratio and peak strain increment ratio. As shown in **Figure 21**, an approximate linear relationship can be found between q_f/p and $-(d\varepsilon_v/d\varepsilon_1)_f$. When the elastic deformation is neglected (i.e., $d\varepsilon_v/d\varepsilon_s = d\varepsilon_v^p/d\varepsilon_s^p$), M (q_f/p) is equal to 1.83 based on the stress-dilatancy equation in a Cam-clay model [22].

6. Constitutive model

6.1 Strength condition

Particle breakage occurs in general sand under higher confining pressure and in weathered granite soil under lower confining pressure [23]. According to the generalized nonlinear strength theory proposed by Yao et al. [24], M_f and M_c are expressed as follows [25, 26]:

$$M_f = M\left(\frac{p}{p_c}\right)^{-n}, \tag{7}$$

$$M_c = M\left(\frac{p}{p_c}\right)^{n}, \tag{8}$$

where M is the stress ratio at critical state, M_f is the stress ratio at shear failure, M_c is the stress ratio at characteristic state point, p_c is the reference breaking stress, and n is the material parameter. Substituting $M_f = q_f/p$ and $M_c = q_c/p$ into Eqs. (7) and (8), respectively, gets the function of q_f and q_c as Eqs. (9) and (10) on the p-q plane, which are the exponential functions:

$$q_f = Mp_c^n p^{1-n}, \tag{9}$$

$$q_c = Mp_c^{-n}p^{1+n}. \tag{10}$$

It can be seen that the product of M_f and M_c is constant, as shown in Eq. (11).

$$M_f M_c = M^2, \tag{11}$$

The stress ratio at shear failure (M_f) and the stress ratio at characteristic state point (M_c) of granite weathered soil were calculated by a large-scale triaxial test under different confining pressures. **Figure 22** shows the shape of M_f and M_c in p-q plane. In this figure, M is the strength line determined by Mohr-Coulomb strength criterion and M_f is the actual peak strength line. From the figure, it is found that the Mohr-Coulomb strength of compacted weathered granite soil under lower confining pressure is smaller than that of the actual strength, while the strength of compacted weathered granite soil under higher confining pressure is higher than that of the actual strength. In other words, By using Mohr-Coulomb criterion, the strength of compacted weathered granite soil under lower confining pressure may be underestimated, while that of compacted weathered granite soil under higher confining pressure may be overestimated. The phenomenon of particle breakage has been found when the triaxial test for weathered granite soil in Ube, Japan, under 10–300 kPa confining pressure has been carried out by Miura et al. [23]. It can be considered that the inapplicability of the Mohr-Coulomb strength criterion for the weathered granite soil is caused by the particle breakage.

By fitting the stress ratio (M, M_f, and M_c) of weathered granite soil as shown in **Figure 9**, it is found that the expression by using Eqs. (12) and (13) is more accurate than by using Eqs. (7) and (8)

$$M_f = M\left(\frac{p}{p_c}\right)^{-4n}, \tag{12}$$

$$M_c = M\left(\frac{p}{p_c}\right)^{n}. \tag{13}$$

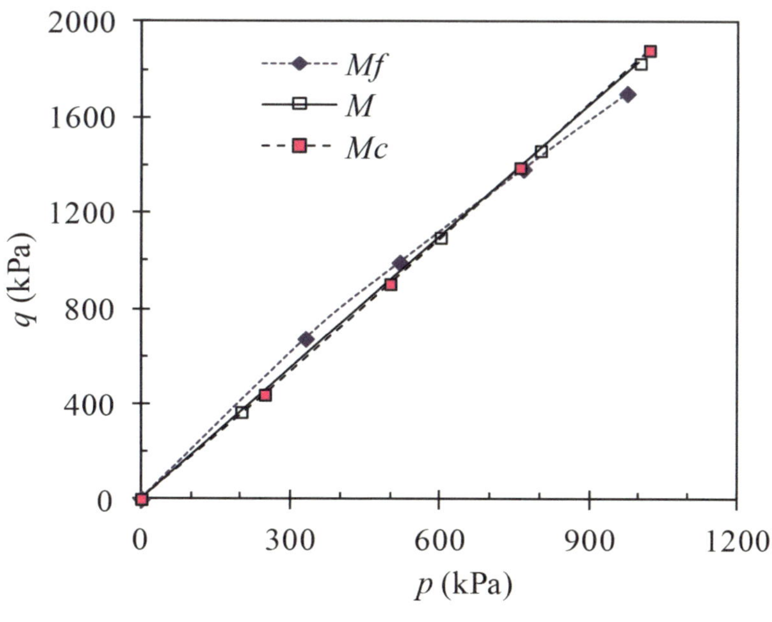

Figure 22.
Shape of M_f *and* M_c *on* p-q *plane for samples.*

According to Eq. (14),

$$M_f M_c = M^2 \left(\frac{p}{p_c}\right)^{-3n}, \tag{14}$$

we can see that the product of M_f and M_c is not a constant, but a power function of p.

6.2 Constitutive model

6.2.1 Yield function

Based on the results of the isotropic consolidation test for Toyoura sand, Nakai [27] has established the relationship between the volumetric strain (including elastic volumetric strain and plastic volumetric strain) and the mean principal stress of sand,

$$\varepsilon_v^e = C_e \left[\left(\frac{p}{p_a}\right)^m - \left(\frac{p_0}{p_a}\right)^m \right], \tag{15}$$

$$\varepsilon_v^p = (C_t - C_e) \left[\left(\frac{p}{p_a}\right)^m - \left(\frac{p_0}{p_a}\right)^m \right], \tag{16}$$

where p_0 is the initial mean principal stress, p_a (=101.3 kPa) is the atmospheric pressure, m is a coefficient for sand, C_t is the compression index, and C_c is the swelling index. Based on the Eqs. (15) and (16), the yield function of weathered granite soil is constructed in this chapter.

When considering the particle breakage of soil, the stress-dilatancy equation in the modified Cambridge model was changed by Yao et al. [28] to the following form:

$$\frac{d\varepsilon_v^p}{d\varepsilon_s^p} = \frac{M_c^2 - \eta^2}{2\eta}, \tag{17}$$

where ε_s^p is the plastic shear strain and η is q/p.

The orthogonality condition is

$$-\frac{d\varepsilon_v^p}{d\varepsilon_s^p} = \frac{dq}{dp}, \tag{18}$$

The expression of the plastic potential function, which is obtained through the differential equation, composed of Eqs. (17) and (18), is expressed as

$$f = (2n+1)\frac{p_c^{2n} q^2}{M^2 p} + p^{2n+1} - p_x^{2n+1} = 0. \tag{19}$$

According to Nakai's research [27], p_x is expressed as

$$p_x = \left(\frac{p_a^m}{C_t - C_e}\varepsilon_v^p + p_0^m\right)^{\frac{1}{m}}. \tag{20}$$

Substituting the Eq. (20) into the Eq. (19), we get

$$f = \frac{C_t - C_e}{p_a^m}\left\{\left[\frac{(2n+1)p_c^{2n}}{M^2}\frac{q^2}{p} + p^{2n+1}\right]^{\frac{m}{2n+1}} - p_0^m\right\} - \varepsilon_v^p = 0. \tag{21}$$

The above expression is the yield function of weathered granite soil. In the function, ε_v^p is an independent variable and can be used as the hardening parameter of the constitutive model of weathered granite soil.

6.2.2 Hardening parameter

On the basis of existing research results, Yao et al. [29] proposed the hardening parameters considering particle breakage as follows:

$$H = \int \frac{M_c^4}{M_f^4}\frac{(M_f^4 - \eta^4)}{(M_c^4 - \eta^4)} d\varepsilon_v^p. \tag{22}$$

Simplifying the differential form of Eq. (21), we obtain

$$d\varepsilon_v^p = \frac{M_f^4}{M_c^4}\frac{(M_c^4 - \eta^4)}{(M_f^4 - \eta^4)} dH. \tag{23}$$

Eq. (23) is used as a hardening parameter to construct the constitutive model of weathered granite soil.

6.2.3 Constitutive relation

By using the theory of plastic displacement potential and the associated flow rule and using the basic assumptions in the Cambridge model, after $d\varepsilon_v^p$ in the yield function, Eq. (21), was replaced by the hardening parameter H considering particle breakage in Eq. (22), we obtain,

$$\frac{\partial f}{\partial p} = \frac{C_t - C_e}{p_a^m} m\left[\frac{(2n+1)p_c^{2n}}{M^2}\frac{q^2}{p} + p^{2n+1}\right]^{\frac{m-2n-1}{2n+1}}\left(p^{2n} - \frac{p_c^{2n}}{M^2}\frac{q^2}{p^2}\right), \tag{24}$$

$$\frac{\partial f}{\partial q} = \frac{C_t - C_e}{p_a^m} m\left[\frac{(2n+1)p_c^{2n}}{M^2}\frac{q^2}{p} + p^{2n+1}\right]^{\frac{m-2n-1}{2n+1}}\left(\frac{2p_c^{2n}}{M^2}\frac{q}{p}\right), \tag{25}$$

$$\frac{\partial f}{\partial H} = -1 \tag{26}$$

According to the definition of generalized deviant stress, the plastic volumetric strain can be expressed as follows by means of a series of transformations of Eqs. (24)–(26):

$$d\varepsilon_v^p = \frac{m(C_t - C_e)}{p_a^m}\frac{M_f^4}{M_c^4}\frac{(M_c^4 - \eta^4)}{(M_f^4 - \eta^4)}\left(dp + \frac{2\eta}{M_c^2 - \eta^2}dq\right)$$
$$\left[\frac{(2n+1)p^{2n+1}}{M_c^2}\eta^2 + p^{2n+1}\right]^{\frac{m-2n-1}{2n+1}}\left(p^{2n} - \frac{p^{2n}}{M_c^2}\eta^2\right). \tag{27}$$

The stress-dilatancy equation is expressed in the following form from Eqs. (17) and (27)

$$d\varepsilon_s^p = \frac{m(C_t - C_e)}{p_a^m}\frac{M_f^4}{M_c^4}\frac{(M_c^2+\eta^2)2\eta}{(M_f^4-\eta^4)}\left(dp + \frac{2\eta}{M_c^2-\eta^2}dq\right)\cdot \left[\frac{(2n+1)p^{2n+1}}{M_c^2}\eta^2 + p^{2n+1}\right]^{\frac{m-2n-1}{2n+1}}\left(p^{2n} - \frac{p^{2n}}{M_c^2}\eta^2\right) \tag{28}$$

Equations (26) and (27) are the plastic strain expression of weathered granite soil considering particle breakage.

In three-dimensional axisymmetry, the following expressions can be obtained from the generalized Hook's law:

$$d\varepsilon_v^e = \frac{3(1-2v)}{E}dp \tag{29}$$

$$d\varepsilon_s^e = \frac{2(1+v)}{3E}dp \tag{30}$$

where E is the elastic modulus and v is the Poisson's ratio.

After taking the derivative of p in Eq. (15), the elastic volumetric strain can be expressed as

$$d\varepsilon_v^e = \frac{mC_e p^{m-1}}{p_a^m}dp \tag{31}$$

Combining Eqs. (29)–(31), the elastic shear strain is obtained in the following form,

$$d\varepsilon_s^e = \frac{2(1+v)mC_e p^{m-1}}{9(1-2v)p_a^m}dp \tag{32}$$

According to elastoplastic mechanics, the total volumetric strain and the total shear strain are, respectively, expressed as

$$d\varepsilon_v = d\varepsilon_v^e + d\varepsilon_v^p = \frac{m(C_t - C_e)}{p_a^m}\frac{M_f^4}{M_c^4}\frac{(M_c^4-\eta^4)}{(M_f^4-\eta^4)}\left(dp + \frac{2\eta}{M_c^2-\eta^2}dq\right)\cdot \left[\frac{(2n+1)p^{2n+1}}{M_c^2}\eta^2 + p^{2n+1}\right]^{\frac{m-2n-1}{2n+1}}\left(p^{2n} - \frac{p^{2n}}{M_c^2}\eta^2\right) + \frac{mC_e p^{m-1}}{p_a^m}dp \tag{33}$$

$$d\varepsilon_s = d\varepsilon_s^e + d\varepsilon_s^p = \frac{m(C_t - C_e)}{p_a^m}\frac{M_f^4}{M_c^4}\frac{(M_c^2+\eta^2)2\eta}{(M_f^4-\eta^4)}\left(dp + \frac{2\eta}{M_c^2-\eta^2}dq\right)\cdot \left[\frac{(2n+1)p^{2n+1}}{M_c^2}\eta^2 + p^{2n+1}\right]^{\frac{m-2n-1}{2n+1}}\left(p^{2n} - \frac{p^{2n}}{M_c^2}\eta^2\right) + \frac{2(1+v)mC_e p^{m-1}}{9(1-2v)p_a^m}dq. \tag{34}$$

The Eqs. (33) and (34) are the constitutive relation of the weathered granite soils considering particle brakeage.

The seven parameters in the constitutive model of weathered granite soil are C_e, C_t, m, M, p_c, n, and ν. Except for Poisson's ratio ν, the other six parameters can be obtained by conventional triaxial test. The value of ν is assumed to be 0.3. C_e, C_t, and m can be obtained by isotropic compression and unloading test, and M, p_c, and n can be obtained by consolidated drained triaxial test.

7. Conclusions

In this chapter, routine physical and mechanics tests and large-scale triaxial tests were conducted to investigate the compaction, bearing, strength and shearing-dilatancy characteristics, and constitutive model of compacted weathered granite soil. The main conclusions obtained from the study in this chapter are summarized as below:

1. With an increase in burying depth of weathered granite, the geological year's parameter (m) decreases by power function, but the geometric progression constants (r) increase by power function. The exponent of power function in this chapter can be used to evaluate the weathering process of weathered granite [11].

2. The particle-breaking characteristics of weathered granite soils are obviously influenced by many factors such as particle gradation, mineral content, blows per layer, and stress level.

3. The pure weathered granite soil shows similar compaction characteristics of sands. The compression and bearing characteristics of weathered granite soil vary significantly when the clay content ratio changes. The weathered granite soil in this chapter can be used as the filling material.

4. The experimental results show that the increase in peak deviatoric stress q_f due to an increase in mean stress p is observed as nonlinear under lower confining pressure because of the existence of particle breakage. It is found that the product of the state stress ratio and peak stress ratio is not a constant but a power function of an average main stress. The research results are helpful to understand the law of the long-term degradation of subgrade performance due to particle breakage of weathered granite soil.

Given that wet conditions may influence the evolution of embankments after construction, future extensive research should be focused on dynamic measurement methods of particle breakage and long-term behavior degradation analysis of subgrade as a result of further weathering and particle breakage [7, 11, 30, 31]. It is very difficult to entirely understand the road performance of weathered granite soil due to its special mechanical properties. Further research needs to be performed for assessing the unique mechanical behaviors of compacted weathered granite soil considering particle breakage and the mechanical properties of different types of weathered granites soils (e.g., coarse-grained weathered granite soil and fine-grained weathered granite soil). Given that measuring particle breakage and its variations during triaxial compression tests are still a challenging task [32], future extensive research should be focused on the dynamic measurement methods, the mathematical description of particle breakage, and the long-term behavior degradation owning to particle breakage [33].

Acknowledgements

The study in this chapter was sponsored by the China Postdoctoral Science Foundation (2016M591044), the National Basic Research Program of China (973 Program, 2014CB047006), the Research Plan of Shanxi Province Department of

Transportation in China (2016-1-7), and the Open Fund of Key Lab of Highway Construction and Maintenance Technology in Loess Region of Shanxi Transportation Research Institute (KLTLR-Y14-17). All the supports are gratefully acknowledged for generously funding this work. The author also thanks Prof. Yangping Yao, from Beihang University, China, for his generous support and assistance.

Conflict of interest

The authors declare no conflict of interests.

Author details

Xirong Niu
Department of Civil Engineering, Shanxi University, Taiyuan, Shanxi, China

*Address all correspondence to: niuxirong@sxu.edu.cn

References

[1] Lu F, Sang L. Petrology. Beijing: Geological Publishing House; 2002. p. 120

[2] Wu N, Zhao C, Hou W. Research on the cause of formation, distribution and engineering characteristics of the granite residual soil. Journal of Pingdingshan Institute of Technology. 2004;**13**:1-4

[3] Writing Committee of Handbook of Engineering Geology (WCHEG). Handbook of Engineering Geology. 4th ed. Beijing: China Building Industries Press; 2007. p. 125

[4] Fukumoto T. A grading equation for decomposed granite soil. Soils and Foundations. 1990;**30**:27-34. DOI: 10.3208/sandf1972.30.27

[5] Yu F. Particle breakage and the drained shear behavior of sands. International Journal of Geomechanics. 2017;**17**:04017041. DOI: 10.1061/(ASCE)GM.1943–5622.0000919

[6] Niu X, Yao Y, Chen Z. The strength and constitutive model of compacted weathered granite soils in Lüliang mountains. Rock and Soil Mechanics. 2017;**38**:2833-2840. DOI: 10.16285/j.rsm.2017.10.008

[7] Konrad JM, Salami Y. Particle breakage in granular materials—A conceptual framework. Canadian Geotechnical Journal. 2018;**55**:710-719. DOI: 10.1139/cgj-2017-0224

[8] Ministry of Transportation of the People's Republic of China (MOTPRC) Test Methods of Soil for Highway Engineering (JTG E40-2007). Beijing: People's Communications Press; 2007

[9] China Electronics Corporation (CEC). Code for Coarse-Grained Soil Tests for Hydropower and Water Conservancy Engineering (DL/T 5356-2006). Beijing: China Power Press; 2006

[10] Geng Y, Yang C, Song B, Wan Y. Post-orogenic granites with an age of 1800 ma in Luliang area, North China Craton: Constraints from isotopic geochronology and geochemistry. Geological Journal of China Universities. 2004;**10**(4):477-487. DOI: 10.3969/j.issn.1006-7493.2004.04.001

[11] Niu X, Yao Y, Sun Y, Luo Z. Weathering process of in situ granite and particle breakage characteristics of compacted weathered granite. Applied Sciences. 2018;**87**:1108. DOI: 10.3390/app8071108

[12] Hardin BO. Crushing of soil particles. Journal of Geotechnical Engineering. 1985;**111**:1177-1192. DOI: 10.1061/(ASCE)0733-9410(1985)111:10(1177)

[13] Leslie DD. Large scale triaxial tests on gravelly soils. In: Proceedings of the 2nd Pan-American Conference on Soil Mechanics and Foundation Engineering; 16-24 July 1963; Rio de Janeiro, Brazil. Vol. 1. São Paulo, Brazil: Associacao Brasileira de Mecanica dos Solos; 1963. pp. 181-202

[14] Leslie DD. Shear Strength of Rockfill, Physical Properties. Sausalito, CA, USA: Engineering Study No. 526; South Pacific Division, Corps of Engineers Laboratory; 1975. p. 124

[15] Marsal RJ. Discussion of shear strength. In: Proceedings of the 6th International Conference on Soil Mechanics and Foundation Engineering; Montreal, QC, Canada. Vol. 3. 1965. pp. 310-316

[16] Lee KL, Farhoomand I. Compressibility and crushing of granular soils in anisotropic triaxial compression. Canadian Geotechnical Journal. 1967;**4**:68-86. DOI: 10.1139/t67-012

[17] Niu X, Sun Y, Yao Y, Han J. Basic properties and weathering process of

strong weathered granite. Japanese Geotechnical Society Special Publication. 2015;**1**:8-33. DOI: 10.3208/jgssp.CPN-07

[18] Lee IK, Coop MR. The intrinsic behaviour of a decomposed granite soil. Geotechnique. 1995;**45**:117-130. DOI: 10.1680/geot.1995.45.1.117

[19] Liu M, Gao Y, Liu H, Chen Y. Large-scale triaxial test study on deformation and strength characteristics of rockfill materials. Chinese Journal of Rock Mechanics and Engineering. 2003;**22**: 1104-1111. DOI: 10.3321/j.issn: 1000-6915.2003.07.011

[20] De-Mello VFB. Reflections on decisions of practical significance to embankment dam construction. Géotechnique. 1977;**27**:281-355. DOI: 10.1680/geot.1977.27.3.281

[21] Vesic AS, Clough GW. Behavior of granular materials under high stresses. Journal of the Soil Mechanics and Foundations Division. 1968;**94**:661-688

[22] Roscoe KH, Burland JB. On the generalized stress-strain behavior of "wet" clay. In: Engineering Plasticity. Cambridge, U.K.: Cambridge Univ. Press; 1968. pp. 535-609

[23] Miura N, O-Hara S. Particle-crushing of a decomposed granite soil under shear stresses. Soils and Foundations. 1979;**19**:1-14. DOI: 10.3208/sandf1972.19.3_1

[24] Yao Y, Lu D, Zhou A. Generalized non-liner strength theory and its transformed stress space. Science in China, Series E Technological Sciences. 2004;**47**:691-709. DOI: 10.1360/04ye0199

[25] Yao Y, Zhou A, Lu D. Extended transformed stress space for geomaterials and its application. Journal of Engineering Mechanics. 2007;**133**: 1115-1123. DOI: 10.1061/(asce)0733-9399(2007)133:10(1115)

[26] Yao Y, Sun D, Matsuoka H. A unified constitutive model for both clay and sand with hardening parameter independent on stress path. Computers and Geotechnics. 2008;**35**:210-222. DOI: 10.1016/j.compgeo.2007.04.003

[27] Nakai T. An isotropic hardening elastoplastic model for sand considering. Soils and Foundations. 1989;**29**:119-137

[28] Yao Y, Matsuoka H, Sun D. A unified elastoplastic model for clay and sand with the SMP criterion. In: Proceedings of the 8th Australia-New Zealand Conference on Geomechanics; Hobart, Australia. Hobart: Australian Geomechanics Society; 1999. pp. 997-1003

[29] Yao Y, Sun D, Luo T, Matsuoka H. A simple 3-D constitutive model for both clay and sand. Chinese Journal of Geotechnical Engineering. 2002;**24**: 240-246. DOI: 10.3321/j.issn: 1000-4548.2002.02.026

[30] Wang W, Coop MR. An investigation of breakage behaviour of single sand particles using a high-speed microscope camera. Géotechnique. 2016; **66**:1-15. DOI: 10.1680/jgeot.15.P.247

[31] Niu X, Xie H, Sun Y, Yao Y. Basic physical properties and mechanical behavior of compacted weathered granite soils. International Journal of Geomechanics. 2017;**17**:04017082. DOI: 10.1061/(ASCE)GM.1943-5622.0000983

[32] Indraratna B, Thakur PK, Vinod JS, Salim W. Semiempirical cyclic densification model for ballast incorporating particle breakage. International Journal of Geomechanics. 2012;**12**:260-271. DOI: 10.1061/(ASCE) GM.1943-5622 .0000135

[33] Chazallon C, Hornych P, Mouhoubi S. Elastoplastic model for the long-term behavior modeling of unbound granular materials in flexible pavements. International Journal of Geomechanics. 2006;**6**:279289. DOI: 10.1061/(asce) 1532-3641(2006)6:4(279)